THEME PARK DESIGN

Behind the Scenes with an Engineer

Steve Alcorn

THEME PARK DESIGN

For information contact :
Steve Alcorn
writing@alcorn.com

Book and Cover design by Dani Alcorn
Cover art by Nick Fang
All photos by Steve Alcorn except where noted
ISBN: 9798652700775

Second Edition: June 2020

Thanks to the employees of Alcorn McBride Inc, who for more than thirty years have devoted themselves to making theme parks reliable, fun and exciting. I'm blessed to have a staff of such special people.

CONTENTS

INTRODUCTION

I'm Steve Alcorn, CEO of Alcorn McBride Inc., a company that engineers equipment for theme parks all over the world. I've been designing theme park attractions and equipment for them since 1982, when I worked on Disney's Epcot Center in Florida, USA. Since then, theme park projects have taken me all over the globe, and my company—Alcorn McBride Inc.—makes the audio, video and control products used in nearly all of the world's major theme parks. It's been great fun bringing hundreds—perhaps thousands—of attractions to life all over the world. Now I'd like to share some of that experience with you.

This is not an engineering textbook, by any means. But it does look behind the scenes from an engineer's perspective.

No book can really tell you how to design a theme park. But this book will introduce you to the people who design theme parks, and help your decide which role you might like to fill.

So who is this book for?

Two sets of people, really. It's aimed at those who are considering a career in any area having to do with themed entertainment. And it's aimed at anyone who has ever gone to a theme park and wondered, "How did they do that?"

This book explores the types of attractions, project stages, and people involved in the theme park world.

By the time you've finished this book you will understand theme parks better as a guest. And for those interested in a career in theme park engineering, you'll have an idea of the different disciplines that you may chose from, and the types of knowledge you'll need to qualify for those positions.

Steve Alcorn at Epcot, 1982

The Accidental Engineer

I ended up a theme park engineer by accident. It all started when my wife was in the third grade. Really.

You see, that's when she decided that when she grew up she was going to design rides for Disneyland. She started with a piece of plywood she covered with little trees and cardboard buildings, making a scale model of the whole park. It even had a tiny little Skyway, with droplets of clay dangling from a copper wire.

I came along later. Much later.

As a college student and inveterate electronic tinkerer, I started a company that made microcomputers before IBM invented something called the PC. Newly married to this woman determined to design rides for Disneyland, the pull of theme parks was strong. Upon graduating she interviewed with exactly one company: Walt Disney Imagineering (then called WED Enterprises). She was hired to help design EPCOT Center, and within two years she was the Electronic Project Engineer on a half dozen attractions. Talk about a trial by fire.

Surrounded by theme park engineers, I was inevitably drawn in. Called in as an outside consultant (a "guru from afar") to help with the particularly complex American Adventure attraction, I set out on a two-week business trip to Florida. Two years later, I came home.

But once you've had a nibble, you're hooked.

And that's how I ended up a theme park engineer by accident.

PART 1
Attractions

What is a Theme Park?

There are many possible definitions of a theme park. Basically, it's a collection of attractions grouped by their motif. For example, one area might be devoted to history, another to nature, and a third to technology.

Walt Disney is usually credited with creating the first theme park, California's Disneyland, because he was dismayed at the low quality of the amusement parks available to him and his young daughter. (An amusement park, unlike a theme park, contains rides, but they are not linked by any theme.)

But there are older examples of theming than Disneyland. Tivoli Gardens in Copenhagen, for example, has an area devoted to cuisines of the world, and another for rides. But its ride area fails the theming test, as there is little relationship between the different rides.

On the other hand, at Disneyland attractions are grouped into Frontierland (the old West), Tomorrowland (the future), Fantasyland (fairy tales) and others. The theming through each land is consistent between rides, shows, restaurants and other attractions.

With the success of Disneyland, other theme parks began to appear all over the world. Many examples in this book are drawn from Disney's Epcot Center in Florida, a park divided into Future World and a World Showcase of different countries' culture. It is similar to what you would find at a World Expo, an annual event hosted in different countries every year. So, in a sense, an expo is also a theme park.

In addition to theme parks, there are also themed attractions, which are smaller, but also consist of a group of related entertainments. A visitor center with rides and shows, or a museum dedicated to a specific area of exploration might qualify.

Epcot Center Construction, July 1982

Rides

When people think of theme parks, the first thing they think of are attractions. Everyone has his or her favorites. When I was a kid it was a toss up whether my favorite was Disneyland's Pirates of the Caribbean or The Haunted Mansion. With delicious anticipation, I'd stand in the queue line soaking up the set design and theming, getting completely immersed in the environment; I would soon eagerly board the ride and try to capture every detail as we rushed through and then, as soon as the ride was over, run around to the front of the building, get back in line and do it again. In retrospect, I now know the thing that made these my favorite attractions was the story they told. I've come to realize that story is the key ingredient in any great themed attraction. If we don't leave our guests with the memory of an emotional journey, we haven't really succeeded in taking them on the ride they paid for.

Even though I was a kid a pretty long time ago—let's just say that back then dinosaur rides involved the real thing—nearly all of the storytelling techniques found in today's theme parks had already been invented. What hadn't been invented are the technologies we use today to tell those stories. Simulators, digital video, wire-guided vehicles, even microcomputers themselves have all come along since I was a kid. It's impossible to imagine building any new attraction without using some—perhaps even all—of those technologies. Yet even as we do, we need to remember story is still king.

Let's look at the different types of experiences available in today's theme parks, and try to understand how they transport us to another reality and keep us coming back.

Roller Coaster, Photo by Paul Brennan

Gravity and Iron Rides

For pure, pulse-quickening power, nothing can really match falling out of your mom's shopping cart. But for those of us too big to fit in that little fold-down seat anymore—and too chicken to take up skydiving or bungee jumping—there are gravity rides. These are the roller coasters, parachute drops, "Big Shots" and anything else that uses the principle what goes up must come down. In the trade, we call them "hard iron" rides because usually the track and its supports are welded steel. For the roller coaster purist, though, old-fashioned wooden scaffolding still holds a certain allure. But the advent of tubular steel rails—often pressurized and monitored to detect fractures—has enabled coasters to do things never imagined with wood: vertical loops, cork screws, and speeds exceeding 100 miles per hour are features of some of today's favorite gravity rides.

Most gravity rides don't tell much of a story, and perhaps that's why I'm not a big fan. But they're a definite lure for a certain type of theme park guest, and some fairly unthemed theme parks have very little except gravity rides.

Dark Rides

Dark rides are, well, rides in the dark. They date back at least to the Coney Island era and the scandalous Tunnel of Love. But today we're likely to think of something a bit more family oriented when we talk about dark rides. The classic dark rides of the Disneyland era are those in Fantasyland: Peter Pan's Flight, Alice in Wonderland, Snow White's Scary Adventures, and, my favorite, Mr.

Toad's Wild Ride. One can also argue some very high-tech modern rides such as Spiderman at Universal Studios Florida and Tower of Terror at Disney's Hollywood Studios are simply elaborate dark rides, but I think that is selling them short. They are really hybrids, and will be discussed in a separate section.

Flintstones Dark Ride, Photo By Wolfgang Eckert

After the initial construction of Fantasyland, Disney upped the ante for dark rides by introducing Pirates of the Caribbean and Haunted Mansion. These two dark rides became the standard by which all others are still judged, because they had three dimensional animation, special effects, synchronized audio throughout the ride, and strong storylines.

For our purposes, dark rides are relatively small, low-throughput rides where individual vehicles—or sometimes strings of vehicles—move through sets and animation that is often illuminated chiefly by ultraviolet ("black") light, making the colors glow vividly. The vehicles may be suspended from overhead as in Peter Pan's Flight, or pulled by a chain drive as in Epcot's defunct World of Motion, or float, but usually they follow a single, wiggly rail, which may ascend—as in Alice in Wonderland—or simply wander all over hell's half acre, as in Mr. Toad, who literally found himself bedeviled in the end.

Boat Rides

It floats. It's a boat. What makes it a ride? Usually a track or a trough that guides the boat through scenery and animation. Some dark rides use boats as a conveyance mechanism, but most boat rides are outdoors.

Europa Park Boat Ride, Photo by Martin Fuhrmann

Boats can be small, such as in the Storybook Land Canal Boats, medium as on the Jungle Cruise, or large as in the Mark Twain paddle-wheeler (yes, it's on tracks). They can also sometimes be guest-piloted, notably the wilderness canoe ride.

Simulators

This was the great technical breakthrough of the 1980s. Used for decades to train military and commercial airline pilots, simulators first made their way into the themed entertainment market at CN Tower in Toronto and, shortly after that, in Disney's collaboration with George Lucas, Star Tours. Simulators allow theme parks to take guests to places that could never be convincingly simulated by other ride mechanisms.

Some simulators place the guests on a moving motion base in a large, open space. An example of this was Universal Studio's Back to the Future, where multiple motion bases operated beneath one of two IMAX dome theaters. (The motion bases are still there, but the ride is now The Simpsons.)

More convincing—to my mind—are simulators such as Star Tours and SeaWorld's Wild Arctic where the guests are completely surrounded by the simulator, allowing no fixed external reference points. Of course, this also means that unless the simulation is nearly perfect, motion sickness is a real possibility. Body Wars at Epcot's Wonders of Life was a good example of this problem. The enclosed environment and rhythmic motion, coupled with discrepancies between the motion profile and the projected scenery made many guests physically ill.

Universal Studios Florida
Earthquake Train Simulator, 1990

One of the things that attracted theme parks to simulators was the ability to re-theme attractions without the capital investment of new props and animation. However, it is often many years between refreshes of the simulator movies.

Lately simulators have become inexpensive enough to find their way into neighborhood malls and family entertainment centers. Ironically, the flexibility of the simulator, which originally attracted the theme parks, now works against them, since once they are ubiquitous, the simulator no longer carries the cachet that made it special in the theme park.

To counter this, theme parks have begun upping the ante, combining simulators with other technologies. The frontrunner in this endeavor is Universal Studios. Their Spiderman ride at Islands of Adventure is widely regarded as the world's greatest theme park attraction. It places an open top motion base on a ride vehicle and moves it through an extremely elaborate dark ride with 3-D projection in each scene. The 3-D animation is perfectly synchronized with the vehicle and motion base, allowing characters to jump from buildings, with their impact "felt" as they land on the vehicle. It's a truly amazing experience.

Virtual Reality

One step beyond simulators, virtual reality eliminates even the moving theater from the ride experience. Don the goggles, and the environment is delivered directly to your pupils. Disney's Research and Development created a sophisticated VR attraction for their DisneyQuest family entertainment centers. Sitting on a bench and using handlebars to guide your motion, you flew your magic carpet though Aladdin's hometown.

Theme parks have struggled with virtual reality. It presents many challenges that make it a problem for their environment. Virtual reality is very low throughput. Because your trip through the alternate reality is generally self-directed, it tends to have a non-linear storyline, or perhaps even no story at all. There is also some evidence that a less than perfect VR experience can lead to alteration in the way the brain responds to visual motion input, after even short periods of exposure; this is a form of brain damage, and needs to be carefully considered before the further proliferation of this technology.

Finally, as personal computers become more powerful, they are able to deliver a wealth of VR experiences at home. Current thinking is that VR represents competition to theme parks, not opportunity.

Carny Rides

Sometimes theme parks use rides found in carnivals and state fairs. These rides typically circle or spin, and occupy a relatively small footprint. Theming brings added value to an otherwise off-the-shelf product. The classic example is Disneyland's popular flying Dumbo ride. In a similar vein, the spinning teacups in Fantasyland are a modification of 1920s boardwalk rides.

It's important when incorporating carny rides into a theme park that there be some "added value". Not just the paint job, but even the purpose of the ride must fit the theme of the area. Otherwise we lose the storyline. Knott's Berry Farm did a relatively poor job of this on their original Fiesta Village area, but additional landscaping, façade work and food service eventually turned the area into a truly Mexican themed land.

Carny Ride, Photo by Marcin

Other Rides

The imagination of theme park designers is nearly unlimited, and there are plenty of rides that don't fit into any of these categories. A plethora may be found at the Legoland parks in Denmark, England, California, Florida, and Germany. Here is a quick sampling:

- Small electric cars kids really drive themselves, on a grid of miniature city streets.

- A motor assisted rope pulley system that allows you to pull yourself and a friend to the top of a tall tower.

🌐 Walk-throughs. These are some of my favorites. Build a ride, then leave out the vehicles. Let the guests walk through the scenes at their own pace. This is a fabulous, inexpensive way to tell a story.

Legoland Electric Cars

Transportation Systems

How to get around the park? Or in some cases between parks. Trains, boats, and monorails are themed attractions in themselves. They do the work of moving guests from one spot to another while entertaining them. Transportation systems have the additional challenge of having to fit in with the theme of every area they pass through, or they disrupt the storyline. That's why Disneyland's monorail couldn't stop in, say, Frontierland. But a train can stop almost anywhere.

A particularly clever use of a train is Universal's Hogwart's Express, which turns a short trip between two sections of the park into a fully themed experience in both directions, as guests experience an adventure on the way from London to Hogwarts—and a different adventure on the return trip.

Monorail, Photo by Paul Brennan

Shows

Most theme parks offer their guests a mix of rides and shows. While rides are the main attraction for teenagers, shows appeal to the younger and older demographics, and provide weary guests a place to rest their feet—or their stomachs.

It's much easier to tell a story in a theater. In fact it's hard NOT to tell a story in a theater. But in theme parks it's important that the theater itself—both inside and out—fit the theme.

There is a temptation, perhaps even a trend, to turn theaters into "black boxes" that can host a variety of shows because they have completely neutral theming. This is a bad idea. Guests are subliminally aware when the story doesn't permeate the entire experience, and this kind of venue tends to leave them feeling like they would have been smarter to just spend a few bucks at the neighborhood Cineplex.

Automated Theaters

Automated theaters represent about half of all theaters in theme park venues. While projection in even your neighborhood cinema is automated to some extent, when we talk about automated theaters we mean theaters where the entire show cycle is automated: entrance doors, exit doors, lighting, and so on.

The actual content of automated theaters varies widely, from a simple video projection to nine-screen Circle-Vision, or a theater with animated figures on a stage, such as Country Bear Jamboree.

One of the best fully automated theaters I've seen is at the Pro Football Hall of Fame in Canton, Ohio. It's the best because of an exceptionally strong storyline.

The show, designed by Edwards Technologies, puts guests in the role of pro football players going to the Super Bowl. Preshow monitors outside the theater let you experience the tailgate party craziness of the parking lot before the big game.

As you enter the theater you're really stepping into a vehicle: it's a rotating platform divided into two halves. In the first half a high-definition film takes you inside the locker room, where the coach is giving his players a pre-game pep talk.

At the end of his talk, the players head out into the corridor that leads to the field, and you begin to rotate 180 degrees. Abstract images of marching players and the echo of their footfalls in the corridors accompany the rotation. The space you arrive in is much larger than the first theater.

A tiny, lighted doorway appears far ahead, then swells in size to occupy a large screen projection as you burst onto the field. The crowd roars in bone-jarring surround sound. And the game begins. In close-up.

Each year NFL films produces a new film for this theater, with spectacular highlights from the most recent Super Bowl. This is a show where not only is story king, it's real.

Epcot Center's American Adventure During Test,
A Fully Automated Show, 1982

Live Theaters

Live theater is the theme park attraction with the longest history, dating back to the Ancient Greeks, or possibly even to cave men wearing funny hats. Traditionally, live theater has been a very manual affair: a guy backstage pulls the levers that control the rigging, and a guy in the rafters works the spotlight. But in recent years, larger stage productions, especially Broadway shows and touring companies, have come to rely on show control systems to handle at least a portion of the cues. Complex lighting, moving set pieces, and special effects all require some level of automation. But the performers still do it the old fashioned way, and the lighting and musical cues are still triggered by a guy—possibly wearing a funny hat—sitting out in the theater someplace.

In theme parks, live shows are likely to be more canned than that. Yes, there's a technical director working one or more "boards" in the audience, but much of the timing for lighting, staging and music is predetermined, and the performers voices may not even be live. In fact the only live thing about a heavily costumed stage show in a theme park may be the guys sweating inside those bulky costumes.

Live Theatre, Wikimedia Images

Hybrid Theaters

Hybrid theaters are mixtures of live performers and highly automated lighting and sets. The dividing line may be thin between these shows and fully automated or live theaters. To me, the distinction is this: if the show would be meaningless without both the live and automated elements combined perfectly, it's a hybrid theater.

A wonderful small hybrid theater is the Aegis show at Nauticus, the National Maritime Center in Norfolk, Virginia. Aegis is the Greek word for shield. It's also the name of the Navy's high-tech protection system used by AEGIS-class destroyers to form a 250-mile-radius shield around a naval battle group.

AEGIS Their at Nauticus, the National Maritime Center

The Aegis Theater combines video, lighting, and effects to simulate a battle situation in the control room of such a ship. Live performers on stage play the role of crew members, and the audience uses voting consoles in the arm of the theater seats to determine the outcome of the show.

Another superb hybrid theater is Mystery Lodge at Knott's Berry Farm in Buena Park California. In this brilliant show designed by Bob Rogers, you are seated in a traditional long house and watch an old Indian summon visions from the fire, shaping the smoke into objects that illustrate his stories. The Indian is a live actor, and the smoke is well, magic. At the end of the show, the live performer evaporates before your eyes.

That's entertainment!

Mystery Lodge, Knott's Berry Farm

Stunt Theaters

Stunt theaters are special forms of hybrid theaters, where live actors not only mix with highly automated props, but do so in potentially dangerous ways. Special life safety issues must be taken into account when dealing with exploding gas, dangerous set-piece movements, and pyrotechnics.

Babelsberg Film Park Stunt Show, Photo by Dirk Brechmann

THEME PARK DESIGN

Astrid Lindgrens Värld, Sweden,
Photo By Efraimstochter

Other Attractions

A full-scale theme park has plenty of things other than major rides and shows. In fact it's mainly this stuff that makes it a "themed" park. What would Adventureland be without a jungle? Or Frontierland without a fort? The landscaping, facades, paving, even the trash receptacles all contribute to the theming. Most of this stuff isn't very interactive, although I have seen talking robotic trash cans.

Here are some of the other themed attractions you'll find in many parks.

Arcades

Arcades are the black hole of theme parks, sucking in endless streams of quarters. They cater to the teenage demographic, and have an extremely low initial cost and almost no operating cost.

The challenge with arcades is to maintain any kind of theming. There are plenty of arcades in neighborhood malls. Finding one in a theme park sometimes seems, well, cheap. To keep the experience special, a smart themed attraction selects games (or custom designs them) to fit into the overall theme, and then adds heavy set decoration to the entire gaming area.

Redemption Games, Photo by Michele Maria

The most successful example of this I've seen was the area outside Star Trek The Experience at the Las Vegas Hilton. It was really a casino, not an arcade, but the only difference is that in a casino you pretend you're not going to lose all your quarters until it actually happens. Everything in the Star Trek Experience casino looked like it came straight off the Enterprise: chrome and black slot machines, barstools, and railings, high-tech wall panels, even overhead laser beams.

Pinball Arcade Games, Photo by Stux

Interactives

Interactive share the theming problem with arcades. An interactive device plopped into the middle of a museum doesn't enhance the venue unless it is completely custom designed—including the console—to specifically fit into the theme. If it's just a computer in a box it may as well be an app you can play at home.

One of the first really successful interactives was SMRT-1, a voice activated robot in Epcot's CommuniCore. SMRT-1 could move, carry on a conversation and play games with ANYONE, responding to "yes", "no", and a few other words. This was a pretty impressive speech recognition feat in 1982.

The best interactive I've seen is the Land the Shuttle simulator at NASA's Space Center Houston. We've all seen flight simulators run on $400 PCs. What Land the Shuttle brings to the game is multiple screens filled with authentic dials and readouts that accurately reproduce the shuttle's control surfaces, and a personal flight instructor named Chet who interacts with you as you try to land that flying brick on an itty-bitty runway. Chet actually walks around among the controls at one point, and appears outside the front window wearing an automobile steering wheel around his neck after particularly rough landings. The media, designed by BRC Imagination Arts, has the player and spectators in stitches from opening until closing every day. And it tells a story.

Land the Shuttle,
Space Center Houston

Exhibits

An exhibit that adequately matches the attraction theming can be just as effective as much more expensive alternatives. As some one who makes his living selling sophisticated control, audio, and video equipment I probably shouldn't admit that well-themed static exhibits are among my favorite attractions. But they are.

*Space Shuttle Exhibit,
Space Center Houston*

An excellent example is King Tut's Tomb at Busch Gardens Florida. It completes the theming of their Montu roller coaster area, but I'd much rather walk through the tomb than throw up on the coaster. Call me silly. Static lighting and continuous run audio impart a suitably spooky atmosphere to a fairly accurate reproduction of the contents of the famous king's burial chamber.

Aside from a short introductory video there is no show. But it works. It tells a story.

Another exhibit area that works is the museum in the China pavilion at Epcot. Whether filled with elaborate animated clocks or terra cotta soldiers, it perfectly complements the China Circle-Vision film and other themed areas of the attraction.

Food Service

While I find food fascinating, not everyone shares my enthusiasm. So theme parks need to go out of their way to make food entertaining. This means either theming the food service, theming the food itself, or both.

Sometimes all that means is the right building. Guests wouldn't buy a hotdog in Tomorrowland if they served it from a plywood hotdog stand. But put it in a chrome flying saucer and call it a Moondog and you can get ten bucks for it.

Epcot's World Showcase is the ultimate in food theming. A continuous international expo where the countries of the world stand side by side, each pavilion serves its indigenous food in appropriate surroundings: Tatami Rooms in Japan, a bistro in France, and so on. There's no question that this theming affects food sales. Many years ago when they installed turn-of-the-century popcorn wagons at Disneyland, things really started popping.

The potential of theming on sales was not lost on the food service industry as a whole. Chuck E. Cheese's Pizza Time Theater was the first to pioneer the highly themed restaurant. It featured a somewhat animated band of animals playing musical instruments and singing to entertain the kids while the pizza cooked. Recent stores have been more modestly themed. Perhaps this is a reaction to the spread of theming to seemingly every neighborhood mall store and restaurant. It becomes increasingly difficult to distinguish yourself. Worst of all, pointless theming doesn't tell a story. It's just, well, theming.

Other heavily themed restaurant chains include The Rainforest Café, Planet Hollywood, and the Hard Rock Café, which has created a worldwide franchise by successfully combining music, food, and rock 'n roll memorabilia.

Mel's Diner at Universal Studios Florida,
Photo by Fabian

Merchandising

Just as the right building can help you sell food, it can also help you sell merchandise. Often the two occupy the same building, and the merchandise is sold after dinner—or after the show or ride—as you "Exit Through Retail".

Appropriate selection of merchandise is critical in a truly themed environment. While you might be able to get away with Tshirts almost anywhere, you cannot sell lava lamps in a medieval castle.

Themed Merchandising, Photo by Sam Chen

Where are Theme Parks
and Who are They For?

Building theme parks is kind of like eating potato chips. Once you've had one, you're almost compelled to move on to the next. Just ask Disney.

Global Expansion

With parks in Florida, California, France, Japan, and China, Disney's clearly placing a bet on the continued financial viability of large parks. And the news is encouraging. Parks on which they spent a lot of money—like Tokyo DisneySea—are extremely popular. The investment is paying off with high attendance and a high rate of returning guests. Other parks, where the budgets were skimpy and the theming was thin, are not faring as well.

Universal Studios also has many parks around the world, and some of the industry's most highly regarded rides.

And then there are rehabs. In the words of Walt Disney, "Disneyland will never be completed. It will continue to grow as long as there is imagination left in the world." That attitude prevails, and ensures that all of the Disney parks will constantly be changing and evolving. It's sometimes painful for those of us in the industry. We work on a project for countless hours, toiling until that fabled 2am on opening day, then watch our creation being demolished a decade later. But that constant flux guarantees the parks will remain fresh and vibrant. And we'll remain employed.

With many other brands also solidly committed to worldwide expansion of their theme parks, the future is likely to hold more and better parks for all of us.

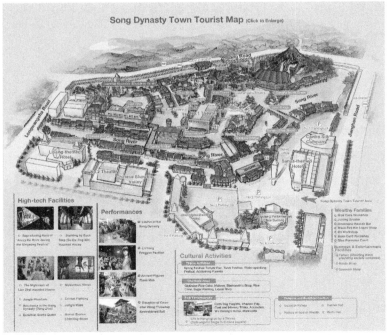

Song Dynasty Town Is the Largest Theme Park in Hangzhou, China

Local Markets

It's unlikely that another park like Epcot will ever be built. Nearly every system, every technology, and every venue was new and revolutionary. Future parks will use more purchased rides, and more guaranteed formulas. But that doesn't mean they need to be boring. Legoland makes up for conservative ride technologies with terrific theming and a precisely targeted appeal: 3 to 12-year-olds and their parents.

Legoland Malaysia, Photo by Fonthip Ward

Six Flags pushes the ride envelope, concentrating on the fastest and most exciting hard iron rides, with minimal theming. The SeaWorld parks and Busch Gardens Tampa focus on animals, and invite guest interaction. Any of these approaches can be successful, if carefully designed to ensure a quality experience for each guest.

Not every park needs to be a giant. Tivoli Gardens has demonstrated this principle for over a hundred years. Smaller regional parks can be successful, even if they are seasonal. Busch Gardens Williamsburg and King's Dominion are examples of successful regional parks.

Except for Disney's parks, European theme parks are all regional. Europa Park in Germany, Liseberg in Sweden, and the UK's Alton Towers are all very popular.

Another type of smaller attraction is the public venue. In Las Vegas you could, at one time, wander beneath a four block long video screen, watch a fountain show or volcanic eruption, and experience a pirate ship battle—including a sinking ship—all for free, from the public sidewalks.

There is still a market for smaller scale, local experiences, and we'll see more of these in the future.

Vertical Markets

Some parks cater to only one specific market segment. Legoland is really just for kids and their parents. Some of the hard iron parks appeal almost exclusively to teenagers.

I believe it's possible to design parks for even narrower vertical markets, provided that the return on investment is properly analyzed. A Native American experience located on Indian land near a major highway in the American West is an obvious example of an untapped market. NASA's Space Center Houston and Kennedy

Space Center are both major visitor centers. Is there room for expansion into a full-scale theme park?

The ultimate vertical markets are corporate ones. Visitor centers, such as the one at Coca-Cola's Atlanta headquarters combine the best elements of themed entertainment with a pitch to buy more product.

Trade shows are also making more and more use of themed entertainment techniques. Automobile shows, in particular, use high budget, state of the art audio, video, and lighting to generate excitement about the manufacturers' latest offerings.

The Fremont Street Experience, Las Vegas, Nevada

LBE

Location Based Entertainment (LBE) is any form of entertainment tied to a particular location other than your home. LBE includes amusement parks, theme parks, ride films, large-scale arcades, bowling alleys, pool halls, water parks, casinos, and multiplex movie theaters. I suppose sporting events should also be included in the list, but for some reason they aren't.

Obviously, LBE has been around for more than a century. But in the 1990s there was a real explosion of LBE startups, focusing on high-end multiplayer interactive and virtual reality computer games. This form of LBE was thought to be the next big thing. Then came the shakeout, and few of those new ventures exist today.

What went wrong?

Many of the new startups were based upon virtual reality or elaborate multiplayer computer simulations. The cost of development for some of these installations exceeded $1000 per square foot, and yet the constrained indoor space in which they operated provided limited revenue generating potential. In short, the bottom line added up to a negative number.

A lot of these facilities didn't properly analyze their capacity. Restricted entryways, slow throughput, and a high spectator to player ratio resulted in many missed revenue forecasts.

Also, most LBE's failed to carry a story line; there was no theming. A roomful of noisy simulators is only appealing to a narrow market segment, and only for limited time. And many LBE's failed to distinguish themselves from one another. One roomful of noisy simulators is a lot like the next. Without clear branding, LBE's failed to attract a repeat audience.

Few LBE's were located in areas with high enough traffic to support their required attendance. Many were in shopping malls. But malls are sparsely populated most weekdays. Teenagers aren't available during school hours to feed dollars into those expensive simulators. Most LBE's generated 80 percent of their income during the 20 percent of the week between Friday afternoon and Sunday morning. Some LBE's located in prime tourist areas fared better.

As home computers became more powerful, much of the LBE experience was available in your family room. Spectacular graphics and accurate real-time simulation were once the exclusive domain of the powerful, expensive computer systems installed in LBE's. Now the most modest desktop computer exceeds their performance.

The interaction between different players was also a unique selling feature of LBE's. But in the world of Internet connectivity, multi-user games involving thousands of people are now commonplace.

But the LBE situation isn't hopeless. One needs to look no further than video games to find an example of an entertainment technology that rose from the ashes.

Video games arrived in force in the early 1980s. Atari, Nintendo, Sega, and a dozen others took off like a rocket, and that first Christmas stores couldn't keep the cartridges on the shelves. But by the late 1980s there was a tremendous shakeout. Public interest waned. But new technologies, PC-based games, and exciting new graphics and storylines revived the industry. Sony's PlayStation and Microsoft's X-box upped the ante. Video games are thriving. It might be possible to orchestrate a similar comeback for LBE's.

To survive, LBE's need to offer a social environment that's not available to someone sitting in his or her family room. A truly successful LBE must involve all of its players in the game. The content is what matters. There needs to be a gathering place where people can meet—perhaps a pre-briefing room—and a cafe, bar or restaurant where they can dawdle afterwards, to talk about their shared experience. You may be able to join thousands of people online from the computer in your family room, but you can't chat with them about the game over drinks afterwards.

A great new development in LBE's is the escape room. These have proliferated widely, and they're fun because each location adds completely different twists and stories based upon that location's owner's own original ideas. That's a marvelous sign of fresh creativity in the themed entertainment industry.

Photo of Ichabod's Escape Rooms,
Courtesy of David Korhonen

FEC's

The Family Entertainment Center (FEC) is a type of LBE that also blossomed in the 1990s, and then experienced a shakeout.

Discovery Zone was the leading FEC. As with most FEC's, the focus of Discovery Zone was on "soft play"—indoor obstacle courses that kids explore, climb through, and slide down.

Discovery Zone was wise to include quiet areas for adults to hang out while their kids played, and a cafe to encourage guests to stay longer. They also provided party rooms for birthdays, and a small arcade for additional revenue.

But the concept wasn't successful because of that old 80/20 problem. 80 percent of revenues were generated during 20 percent of the week between Friday afternoon and Sunday morning. There just weren't enough birthday parties to pay the bills.

Today only the largest and best located FEC's still exist. The FEC is still a concept that bears watching. People are willing to spend a lot of money on their kids, and the right mix of story, environment, and revenue model might still lead to a successful concept.

Discovery Zone

PART 2

Project Stages

Concept

The design of every theme park attraction begins with a Blue Sky phase. This is where the Creative Team sits around a large table and brainstorms new ideas for the attraction.

Most attractions start out with a completely impossible idea, either because it can't be done or can't be done at a reasonable cost. So the design of attractions ends up being a negotiation process between the Creative Team, the Engineers, and the Estimators who are caught in the middle.

Eventually they will all agree on a design that is achievable at a price within the budget. At least, they think they have agreed. Then the engineers go off and design something that is nothing like the Creative Teams imagined. At the same time the Creative Team starts trying to slip back in all the impossible stuff that they previously agreed to take out. And the estimators keep telling both groups not to spend any more money.

It's an iterative process.

Eventually time, money, patience or all three runs out and the attraction opens to the public. Then the Creative Team studies the public's reaction to their creation and comes up with a whole bunch of new ideas for improvements. The engineers now change to the graveyard shift, and try to figure out how to shoehorn in all this new stuff, without exceeding the original budget and without impacting the next day's operation of the attraction.

As you can see, the Blue Sky process never really stops.

Walt Disney once said that as long as there was imagination in the hearts of men, Disneyland would never be finished. I'm not sure whether the irony was intended.

Epcot Calendar, 1982

Who Are Your Guests?

In designing any new park or attraction, the primary consideration is demographics. It means your guests' "background." Here are just a few examples, in order of importance:

- Location of Residence
- Age
- Family Income
- Educational Level
- Race
- Gender
- Religion

There are almost as many types of potential theme park guests as there are types of people in the world. I say "almost" because we can immediately rule out a few categories. Yanamamo tribesmen from the jungles of South America are not very likely to visit our theme park. That seems obvious; I'm instinctively applying selection criteria to narrow and define my audience.

You may already think you know who goes to theme parks: middle class white families from large metropolitan areas. Statistically, if that's your guess, in the US you're right.

There are a few problems with such a simple-minded analysis.

First of all, parks outside the US are the major market for theme park growth. Soon there will be many times the current number of theme parks in China alone.

Second, we need to define what we mean when we say "theme park". To me, a theme park is really any place that offers "dedicated venue" themed attractions. Dedicated venue means that a facility is always used for the same thing. A neighborhood multiplex cinema does not qualify. But a theater that always shows the same film is a dedicated venue. That means that all of the following are themed attractions:

- Museums
- Visitor Centers
- Corporate Communications Centers
- Broadcasting Studio Visitor Tours
- Themed Shopping Malls
- Themed Restaurants
- and many more

Clearly, each of these categories attracts a very different demographic. As designers we need to understand whom we are designing for. A military history museum has completely different attractions and a whole different level of signage than, say, a kids' science exploration center.

A second problem with assuming that our guests are all white middle class families is that it isn't a very accurate descriptor. A park full of roller coasters will appeal to teenagers, but not their parents or younger siblings. Legoland is for the preteen crowd. And Epcot attracts the young and old, but leaves many teenagers unimpressed. As theme park designers we need to understand whether putting a roller coaster in the middle of Epcot will bring in more teenagers or just turn off our current guests.

It's easy to see that demographics will impact our designs in many ways. Never lose sight of this core principle of theme park design: who are your guests?

Florida Museum of Natural History, Gainesville, FL

Launch

It may seem obvious, but the first step in the creation of a new theme park attraction is for someone in management to decide they need a new attraction. That seems basic, but there are many considerations involved. What is the park's current attendance? What will its attendance be if no new attractions are added? How well are the old attractions holding up, both from an artistic and maintenance standpoint? Is there sufficient money or financing available to build a new attraction? What sort of attraction would bring the kind of people the park is looking for?

To some extent the public actually determines this process by their preferences. But it is management's reaction to public preference that initiates the whole process. It would be rare management that went to the Creative Team and said, "We need a new attraction. Come up with something."

A more likely approach is for management to tell the Creative Team, "We need an attraction that appeals to the six to ten-year-old age group," or, "It's too hot at our theme park during the summer. We need an attraction that cools people off by getting them wet."

The Creative Team can now nurture this seed of an idea until it grows up to be a Sequoia. Or a zucchini.

Illustration by Biljana Jovanovic

Blue Sky

There's an interaction with management at this point because the Creative Team won't fully develop any idea until management has approved it. Instead, the Creative Team throws 10 or 15 quick ideas at management to see which ones "stick to wall". Sometimes sticking to the wall is literal. Ideas are posted on bulletin boards and those passing by give their gut reaction.

Blue Sky can also involve the technical disciplines. It would be a foolish management team that embarked on their favorite creative idea without checking with the engineers to get a general idea of how much it will cost. It's all well and good to suggest taking guests to the moon as a big draw for your space theme park. But if the art directors intend to REALLY take people to the actual moon it could get rather expensive. Obviously, that's an exaggeration. But it's easy for engineers to estimate budget for a project based upon its creative content.

One reason it's easy is that no one expects them to be right at this point. If they're smart, they'll inflate the cost by a factor of three, because we all know that the Creative Team will get lots of expensive ideas for improvements as we go along. Still, inflated or not, these numbers are useful for comparing the relative costs of different proposals.

Estimating (and Repeat)

This is where the trouble starts. What if management just LOVES that go to the moon idea, but doesn't love the engineer's estimate that goes along with it?

Re-estimate it, obviously.

Photo by Steve Buissinne

This is where the professional bean counters come in and try to figure out how to get the engineers to make it cheaper. Typically, they take a whole bunch of appealing stuff out of the original idea, and try to do what's left in the least expensive—and lowest quality —manner. The unfortunate results of this process are:

Management becomes convinced that the project can be done economically.

The Creative Team realizes that they're going to end up with an awful attraction unless they can find a way to slip back in all the expensive stuff that was taken out to save money.

The Maintenance guys realize that they're going to end up with an attraction built on such a shoestring budget that they won't be able to keep it running.

Later on, the final results are:

Management is furious when they discover that the attraction ends up costing just what the engineers predicted in the first place; that figure was three times what they wanted to spend.

Why Story Is King

If story is king, then even at the beginning of Blue Sky there must be story.

Let's take an example. A roller coaster careens through a darkened room over a faintly illuminated cityscape. Enthralling? Not really, there's no story.

Take two. A rock band is late for a concert at the Hollywood Bowl. They invite you to hop in their limo and go careening through the Hollywood Hills and all around the L.A. freeway system to make it on time. That's the story behind Disney's Rock 'n' Roller Coaster, and it works.

How about this one: you climb aboard a BART subway train. It pulls out of the station, then begins to shake as an earthquake strikes. Fires erupt, and a flood comes cascading down the tunnel, extinguishing the flames and splashing over the train. Exciting? I guess. But not completely fulfilling. Why? They forgot to tell us why we were getting on the train, where it was going, and what our mission was. The name of this attraction at Universal Studios Florida was Earthquake: The Big One, so we knew what to expect when we got on. But there was no underlying story to get us involved.

Here are two more real ones, one that doesn't work, one that does:

A boat glides through a dark tunnel. It passes a volcano, people at a bazaar trying to sell us things, Mayan ruins, dancing dolls with colorful costumes, and fiber-optic fireworks.

A boat glides through a dark tunnel. It passes a ship full of pirates and a fort. A battle is underway. Cannon balls whiz overhead, and explosions dot the water. Farther along the pirates have seized the village and are setting fire to the buildings. As we barely escape from the burning timbers we see prisoners still trapped in the jail, trying to lure a dog into bringing them the keys to their cell.

Which ride has a story, the original Mexico pavilion at Epcot or Pirates of the Caribbean?

It's not that hard from the outset to make sure that a ride has a story, which makes it surprising that so many rides don't have one. But a lot of them don't. I'm not talking about rides in amusement parks. I'm talking about rides in theme parks. In amusement parks when we see an iron roller coaster we expect to be tossed around a little; we don't expect a story. That's why it's called an amusement park, not a theme park. But at a theme park our expectations are higher. Do the Batman or Superman roller coasters at Six Flags theme parks tell a story, or are they simply themed façades accompanying an unthemed ride?

Sometimes there might be a story there, but it isn't intelligibly conveyed to the riders. At the Journey to Atlantis flume ride at SeaWorld Orlando, preshow monitors show news broadcasts and interviews related to the reappearance of the lost continent of Atlantis. A Greek fishermen is involved, and a statue of a sea horse. The audio is usually intelligible, but the set up doesn't ever give us a mission. Once on the ride the audio becomes unintelligible, and animated figures and props—presumably there to convey a story—pass by so quickly that they can't be perceived. There's some kind of woman or witch in a very bad mood, and a reappearance of the sea horse. Then we go down a really nice drop, get soaking wet, creep back an upramp, and get one final surprise before unloading. It's not a bad ride, but it's incomprehensible.

Sometimes the story is just too complicated for the ride. The Lord of the Rings makes a great book and movie trilogy, but would it make a good ride? Of course not. Rides with more complicated storylines are often best implemented using simulators. Here it is customary to have a narrator—often the driver—who can summarize the adventure as it proceeds. And since simulator rides can be as long as ten minutes, there's more opportunity to convey the story.

Conversely, short rides need simple plots. You step into a basket, are hauled to the top of a tower, and dropped. Or... you go to a creepy old hotel where guests mysteriously vanished in an elevator years before... as you enter the darkened elevator shaft you suddenly feel yourself falling. Knott's Berry Farm's Parachute Drop or Disney's Tower of Terror: which is the better ride? Tower of Terror. Of course, it cost 50 times more.

The year after Tower of Terror opened, it was updated and re-advertised as Tower of Terror 2. The new version dropped guests twice. The next year they added even more drops, and then more. So now that Tower of Terror drops you four or five times for no particular reason, is it a better ride? That's a tough call. It's more exciting, however the story suffers. But it does allow the marketing folks to advertise the new ride profile each year.

A particularly effective mechanism for storytelling is the old-fashioned dark ride. Here a simple vehicle moves along a track—sometimes level sometimes with elevation changes—traveling from scene to scene and telling a linear story. The dark interior allows ultraviolet lighting to focus your guests' attention on the elements most important to the story. Still, one needs to be careful not to try to tell too complex a story. We all know the story of Alice in Wonderland. Without that background the ride at Disneyland would be nearly incomprehensible. But because of that shared background, the audience can relive the book or Disney movie without confusion. This foreknowledge of your guests' background is essential to a successful ride.

Fitting Story to Audience

It's essential to know your audience when designing an attraction. This process of evaluating the audience begins almost from the first moment of Blue Sky and doesn't end until the concept moves from Art Direction to Engineering. Even then the mechanical or ergonomic design of the attraction may be influenced by its anticipated guests. For example, in Europe it's okay to make guests climb stairs or jump off of slowly moving vehicles. In America it's not.

Photo by Gerd Altmann

Let's look at how the designers of some other attractions targeted their audiences in order to see how we should proceed.

When the Las Vegas Hilton decided to install the Star Trek experience, they made a calculated decision to recreate the Starship Enterprise from the television series Star Trek: The Next Generation. They didn't use starships from the classic series of the late '60s or from more recent sequels, such as Star Trek Voyager. Why did they make this calculated decision?

It's because they knew their audience. The majority of people likely to visit the Star Trek Experience (and then spend time at the slot machines in the adjacent casino) were from an age group that would've watched Next Generation on television, but perhaps not the classic television show (except maybe in reruns).

When Legoland built their theme park in Carlsbad California they wanted an attraction where kids would be able to drive cars. There are lots of motorized go-cart racing places around the country, but they appeal more to teenagers than to Lego's target audience of preteens. Also, Lego didn't want anything so environmentally unfriendly and noisy. They wanted something more in keeping with the theme of their park: the Lego brand. So they decided to use electric cars.

The problem with electric cars is that they don't accelerate very quickly or go very fast (unless you have a Tesla). In a world where every third television show ends with a car chase, electric cars are about as exciting as watching finger-paint dry. Lego had to figure out how to make electric cars interesting to kids. The solution is in the story. Lego created a kid-size grid of streets, laid out like the intersections in a real city, complete with stop signs and traffic lights. The idea was for kids to drive around the miniature city, following the same traffic laws that their parents have to follow. Clearly this was something that would appeal to the imagination of an 8-year-old if—and it's a big if—you can get him to do it.

To accomplish this they had to figure out a way to complete the story. The solution was to start the experience off with an instructional video that would teach kids how to observe the rules of the road: obey the signs and lights, use hand signals when turning, and be courteous to other drivers.

Another problem was getting kids to leave the vehicles once their time was up. They found the perfect solution in the completion of the story. At the exit of the attraction the kids are awarded Lego driving licenses.

The result: a simple ride becomes a complete experience though the use of story. A story designed specifically to appeal to the target audience: preteens.

A Dose of Reality

Some of the best amusement rides started out as impossible ideas. As engineers we were called in to help design the control system for the vehicles on the Spiderman ride at Islands of Adventure. Here's the challenge we were facing:

The vehicle moves though eleven scenes, as the onboard motion base pitches and rolls the cabin. Both the vehicle's and the cabin's motion must be precisely synchronized with two off-vehicle projectors that provide a 3D image. The projection combines a portion of the cityscape background with animated characters who appear to swing through the air and even jump on your vehicle.

Synchronization is essential. When Spiderman lands on the front of your vehicle, you must feel the impact. And because your perspective changes as you move through the scene, buildings in the distance must change size at you approach them, and change perspective as you rotate.

The left and right projections must be precisely synchronized to both the vehicle and each other. If 3-D films are even one frame (a tiny fraction of a second) off, the effect is ruined. Because it may take longer to load guests into a vehicle than we expect, we can never be sure when the vehicle will reach a given scene. So we have to stop and start the projection as needed.

Vehicles are mechanical. They don't always go at the speed we command, so we have to speed them up or slow them down to get to the right spot at the right time. But that will affect their synchronization along the way.

Clearly the whole thing is impossible. We even said that while we were working on it. The synchronization would never be good enough to be convincing.

Today it's operating as the world's greatest amusement ride.

Okay. So the engineers aren't always right.

Photo by Gerd Altmann

Practicality

But there are lots of things that can doom a themed attraction. Just because it's possible doesn't mean it's practical. We can send a man to the moon, but not an audience. At the circus they shoot people out of a cannon, but the average guest wouldn't enjoy it. And while I might enjoy an hour lecture about Estonian history, most guests would pay double just to get out.

We also can't afford anything that takes a long time to set up. Physically demanding attractions are pretty much out, as they greatly limit your audience. Dangerous attractions—like walking through a cave filled with recently fed lions—aren't a good idea. And attractions with slow logistics should be avoided—for example, loading guests onto tiny individual rafts and having them pole it across a river isn't very practical.

Maintainability

There's no point in building an attraction if we can't keep it operating. Nearly all rides and automated theaters involve mechanical equipment. This is usually where the trouble starts. It's relatively easy to design mechanisms that can perform a function once. It's much harder to design a mechanism that can repeat that same function hour after hour, day after day. And theme park mechanical systems take a lot of abuse, so they have to be sturdy.

One of the most fun rides ever was the Flying Saucers at Disneyland. One or two guests would board each of a couple dozen small flying saucers. They looked something like yellow inner tubes.

The saucers sat on a large blue surface covered with tiny round flaps. When the ride was started the surface was pressurized with air. The blue flaps held the air in until a flying saucer moved over them. Then fingers on the bottom of the saucer pressed the flaps open, allowing the air to lift the saucer off the ground a few inches. By leaning one direction or another you could cause your saucer to glide around the ride area, bumping into other saucers, rebounding, or just speeding from one side to the other.

It was great fun.

It was also a maintenance nightmare.

There were tens of thousands of those little flaps. If only a few got stuck open there wasn't enough air to hold the saucers up. And they were incredibly difficult to fix. In five years the ride was gone, a victim of its poor reliability.

In 2012 a similar ride, Luigi's Flying Tires, was installed at Disney's California Adventure. It only lasted three years!

Maintenance

Computers present a particular challenge for theme parks. While electronics are one of the most reliable components of a park, the mechanical systems that interact with them are notoriously problematical. You can rule out mice and keyboards right from the start. A theme park audience can destroy them in a day. But even some of the alternatives can be trouble.

At Space Center Houston we installed 24 interactive games about space. For example, one was called Orbital Rendezvous. It allowed you to experiment with the space shuttle's rocket engines, to see how orbital velocity affected orbital distance rather than speed. The goal was to dock with a space station.

The interface mechanism for Orbital Rendezvous was a touch screen. These have the advantage of no moving parts. There are many different touch screen technologies that work fine in an office environment. But in a theme park, touch screens take a severe beating. Only the optical break beam ones completely avoid wear and tear. These weren't that kind. With school kids jabbing their fingers at them all day long they lasted less than a month. At over a thousand US dollars each, it was a good thing the sponsor of this attraction was also providing us with free touch screens.

Budget

Unlike those touch screens, most things aren't free. As theme park engineers we constantly face design, installation and maintenance budget constraints. There is a trade-off between the cost of choosing a technology that won't break and the cost of repairing a technology that will. Most engineers—and certainly all maintenance people—would prefer us to choose the former. But in the real world we often end up designing attractions to fit within the installation budget. The maintenance budget comes out of a completely different pocket.

Worse yet, it's easy for Art Direction to conceive of an attraction that we simply can't afford to build, no matter what technology we use.

I'm thinking our "instantaneous transportation back to medieval times" may fall into this category. Has anybody got any bright ideas yet?

Photo by Chris S

Throughput

Just because we can build an attraction doesn't mean that we should. What if not enough people could enjoy an attraction to justify its expense? Virtual reality is notorious for having this problem. A magic carpet simulator ride allowed you to don a headset, sit on a control seat, and fly around Aladdin's home town. It was an amazing experience. But each simulator could accommodate only one guest for about five minutes. Even with several simulators, only a few hundred guests could try it each day. That's not practical in a theme park that can accommodate 50,000 people.

Photo By Pankgraf

Safety

Some ride concepts are just too dangerous. One way of getting our guests instantaneously transported back to medieval times would be to drop them through a trap door. But even on a theater stage, with trained professionals who are expecting a trap door to open, performers are often hurt.

While state and federal regulation of thrill ride safety has been increasing, the safety of roller coasters and other thrill rides has historically been pretty good. You've got a lot higher chance of being injured in a traffic accident on the way to the theme park than you do while you're in the theme park.

Most theme park injuries occur not on thrill rides, but in more mundane ways. Leaving children untended to crawl out of vehicles or fall into water are two of the biggest causes of injury and death.

As theme park engineers we constantly need to keep in mind all factors that could affect the safety of a ride. And if the fundamental concept of the ride itself is unsafe, it is our responsibility to point that out.

Accessibility

The Americans with Disabilities Act (ADA) legislation and similar laws in other countries have mandated a broad range of accommodations for people with mobility, vision, and hearing disabilities.

This has had a major impact on the design of theme park attractions—and their cost. Now theaters must provide captioning for the hearing-impaired, some signage must be presented in Braille, and most attractions must accommodate people in wheelchairs.

While no one has yet been forced to accommodate wheelchairs on a roller coaster, it's impossible to overstate the affect the ADA has had on the themed entertainment industry. We need to evaluate every attraction for compliance with this law.

Photo by Mabel Amber

Linda Alcorn, Show Control Engineer for Epcot, 1982
(Note the Complete Absence of Electronics.)

Design

At some point, someone makes the decision to actually build something. A project may have languished in creative development for years, waiting for a champion in management to back it. Or it may have been fast-tracked to tie in with an upcoming and—hopefully—blockbuster movie release. In either event, we've now got the bulk of the creative design done, funded the "hard" design, and turned on the guys who really know how to spend money.

Art Direction

We've seen that Art Direction begins during Blue Sky. It doesn't end until the last roll of wallpaper is glued on. Depending upon the budget—and the Art Director—the results (and the cost) can vary widely. But a great Art Director can be the difference between a lackluster attraction and a truly great one.

Want that special paint that doesn't fade in the sun? Synthetic fabric not realistic enough for your animatronics wardrobe? Oh, now that we're using real silk, we have to fireproof it. And of course fake rocks cost more than the real thing.

A big part of the budget goes to those little details that can be so important.

Or not. When Thomas Jefferson and Ben Franklin rise on stage in American Adventure, I doubt anybody notices that all those crumpled up drafts of the Declaration of Independence are actual drafts of the Declaration of Independence!

But it makes a good story.

Epcot Canada Pavilion, June 1982

Facilities Design

The largest single consumer of budget in a themed attraction (unless it's a rehab of an existing attraction, or an outdoor ride) is almost invariably the building itself. Site work, infrastructure (utilities and such), steel, and concrete cost a heap of money—as much as $50 million or more for a complex attraction. Then there are interior walls, finishing, flooring, paint, electrical, plumbing, lighting, emergency lighting, and so on.

The point is, all of this stuff has to be designed sometime by somebody. Well now is the time; and the architects, architectural engineers, and facilities engineers are the somebodies. It's not uncommon for the architectural design alone to cost several million dollars. And that's just the paperwork.

Think about it: a 160-foot self-supporting geodesic sphere anchored by three pillars, with a few million pounds of ride in it. Got one of those in your neighborhood? Didn't think so. But somebody had to make Spaceship Earth stand up, even during hurricanes. That costs money.

You might assume that the show and ride had to be defined before a building could be designed to house them. That's not really the case. The process is somewhat parallel, and major requirements of the interior design do impact the exterior structure. But more often the building constrains the show or ride. That's because construction rushes ahead to stay on the immutable, omnipotent schedule, even while the attraction is being designed and redesigned.

To give you an idea of just how much can change during a project, look at American Adventure. This Epcot pavilion features a theatrical show with a bunch of fourteen-foot-high lifts that carry animated figures from a pit up to stage level. Because different figures must appear in the same spot at various times, the lifts must be rearranged during the show. This is accomplished using a 400,000-pound carriage containing the lifts. Noiselessly, it rolls beneath the audience, indexing to the correct spot for each scene. Obviously this had a major impact on the building. And yet, when construction of the building began it was believed that the lifts would be mounted on a giant turntable!

WED Enterprises Engineering Offices, 1982

Ride Design

Q: Who is the first passenger on a theme park ride?

A: A sandbag.

There are a number of reasons for this. Sandbags have very low health insurance costs. Sandbags look better wearing shorts than the average theme park guest. And sandbags can't sue you.

There are multiple steps to designing a ride system, ranging from turning the creative concept into a working mechanical design, and then figuring out how to control it.

This last step is done by the Ride Control Engineer.

There are many different types of rides to control, and more new ones being invented every year. And just as there are many different types of vehicles, there are many different ways of controlling them. Some require far more sophisticated control systems than others. And some require far more safety analysis.

We'll discuss the Ride Control Engineer in the people section, and all those vehicle types and the hardware used to control them in the equipment section.

Universal Studios Florida Earthquake Train

Show Control Design

As the Show Control Engineer you are often the peacemaker:

You assure the manic-depressive Audio Engineer that the sound can be dynamically controlled to counter the damage wrought in the studio by the tone-deaf sound designer. "Sure, we'll just fade it as the vehicle passes through the scene."

You soothe the hypertensive Special Effects Designer who has just discovered that if the lava pump is activated for more than two seconds it explodes. "We'll just cycle it intermittently from the show control system, one second at a time."

You save the Set Designer from disgrace. "We'll just stop that periscope before it harpoons the ceiling tiles."

Most of all, you are the great communicator.

You make sure the audio engineer knows you are controlling his equipment using contact closures, not a battery from a 1947 Packard.

You make sure the contractor pulls UL approved wire, and keeps it away from that Tesla coil in the preshow.

And you make sure the Project Manager knows you need to start testing your system well before opening day. This last request won't do you any good, but it was worth a try. And you can always use the Project Manager to test that Tesla coil.

Universal Studios Florida,
Jaws Ride Show Action Equipment

Audio/Video Design

The job of adjusting the audio is very subjective, and there are probably as many different opinions on the best settings as there are A/V Engineers on the job.

Some theme parks do the actual audio mix in the theater. This means the master material is actually altered to fit the acoustic performance of the theater, and rerecorded to incorporate these adjustments.

There was a time when theme park audio was expected to be pretty poor. Of course, there was also a time when if you weren't near the stage at a rock concert you couldn't tell what band was playing.

The public's expectations for audio quality in public spaces increased dramatically during the 1970s. By the time that Epcot Center opened in 1982, theme park audio was among the best anywhere. Every year theme park audio becomes more integral to the overall experience.

Audio/Video Cabinets, Mob Museum,
Las Vegas, Nevada

Lighting Design

Lighting is both an artistic and technical endeavor. The selection of fixtures, colors, placement, and transitions all come straight from the theater.

Not the theme park theater. I'm talking about Broadway.

Many theme park lighting designers have worked on Broadway, and most would say the artistic design portion of their job is what they like to do best. But there's no question there's a technical component to the job, and far more so than on the stage.

That's because the average theme park attraction lasts a lot longer than the average stage production, and probably has a much higher budget. Some of that money is spent on making sure the lighting system will be both robust and economical to maintain.

Lighting Test and Adjust,
Epcot's American Adventure, 1982

Lighting designers in theme parks also have the luxury of a year or two of design time, rather than the tightly compressed schedules found in most stage productions. They are also free from the weight and size restrictions that restrain traveling productions, and aren't limited by the capabilities of an existing facility.

That's the big difference. Theme Park Lighting Designers usually have input into the physical design of the spaces they are to light, the location of structures to hold their lighting fixtures, and the amount of power available to feed them. This absence of constraints means that doing lighting design for theme parks can be a lot of fun.

On the other hand, Theme Park Lighting Designers must do the most important part of their job—the field work—in the same incredibly compressed schedule that confronts the other engineering disciplines. This means long nights just before opening day, and competing with the construction trades and other engineers for access to the facility.

Special Effects Design

The Special Effects Designer probably isn't an engineer. He's more likely a guy who grew up reading surplus catalogs, disassembling firecrackers, and dyeing his sister's pigtails green using his junior chemistry kit. Nevertheless, if his pants are too short, he wears a pocket protector, or eats sufficient quantities of Twinkies we'll award him an honorary engineering credential and include him here.

A lot of what Special Effects Designers do is engineering. A lot more is trial and error. Their methods are not so rigorous, nor their results as predictable as we engineers might like. But their designs may incorporate electronics, chemistry, lighting or mechanical systems, so a certain amount of technical knowledge is a must.

The Elaborate Special Effects Set at the Heart of World of Motion During Installation, (Soon To Be Filled With Fog and Lasers), Epcot 1982

Special effects are normally prototyped long before the attraction is built, because they must measure up to the dreams of the Art Director. Among the minor miracles they're asked to perform: lava that bubbles endlessly, cigars that smoke day after day, and buildings that burn forever.

Gee, that's all pretty hot stuff.

But they also might need to come up with: blocks of ice that don't melt, rain that falls ceaselessly, or a ceiling that simulates the roiling surface of the ocean.

The Special Effects Designer's bag of tricks is huge. Effects might incorporate projection, sounds, optics, mechanics, and even smell cannons. These are devices that release a small amount of scent, usually triggered by the passing of a ride vehicle. Popular scents include hot rocks, roses, and stinkbugs. My favorite was called Troll Crotch.

Construction

Once the design is well under way (but long before it is complete) we need to begin construction. And that starts literally from the ground up.

American Adventure, Epcot, June 1982

Site Work

The first step in building any theme park is the site work. This begins with clearing the land, and then continues through all manner of terraforming. At Walt Disney World this meant draining a swamp and the construction of 47 miles of canal.

Epcot Site Work, 1982

While draining wetlands is no longer ecologically acceptable, huge amounts of dirt still need to be moved around to construct most theme parks. Theme parks with elevation changes are visually more interesting than flat theme parks; so most designers take advantage of any natural topography to create an interesting landscape. A great example of using the natural topography is at Legoland California, where the natural hillsides above the Pacific Ocean form an interesting variety of hills and lakes, making a perfect theme park setting.

Even when the land is completely flat extensive earthmoving is sometimes needed. In Florida the porous, sandy soil won't support the weight of theme park-sized buildings. Prior to the start of construction the ground must be "pre-charged" to compact the soil and drive out groundwater. You can see this taking place wherever huge mounds of dirt are piled at the future site of buildings.

Sometimes the earth isn't so much "moved" as it is "delivered". Tokyo Disneyland and Tokyo DisneySea are constructed on fill dumped into Tokyo Bay. Before 1980, the site of Cinderella Castle would have been better suited to fishing.

Not that you'd want to eat any fish that came from Tokyo Bay.

Transportation

Even before earthmoving begins, thought must be given to how a theme park's transportation systems will work. In the United States, most theme park visitors arrive by car. This means massive parking lots must be near the theme park entrance, or near transportation systems that lead to the theme park entrance.

At Walt Disney World, Walt didn't want the reality of the everyday world of parking lots and fast food restaurants to impinge upon the Magic Kingdom, the way they had at the original Disneyland. So he purchased 25,000 acres in Central Florida. This was enough land to relegate the fast food restaurants to the distant fringes of the property (although lately a few McDonald's have found their way into guest areas).

He also located the main parking lot for the Magic Kingdom on the other side of the Seven Seas Lagoon. Guests arrive at the Magic Kingdom by either paddle-wheeler or monorail, so their themed experience begins even before they enter Main Street.

Visitors to theme parks outside the United States are more likely to arrive by mass transportation. While there is still a massive parking lot at Disneyland Paris, the train station was constructed right outside the entrance, providing easy access from anywhere in Europe.

Epcot Sitework and Monorail Beam, 1981

Transportation between attractions is important, too. Walt Disney first recognized this need with the horse drawn carriages on

Main Street. On a larger scale, double-decker buses and ferryboats transported guests around Epcot's World Showcase.

Even when parks are intended primarily for pedestrian traffic, they must still be constructed so that during the night maintenance personnel can maneuver trucks, cherry pickers and other service vehicles within the park.

Most parks also have a periphery road that runs around the outside of the attractions, providing discrete daytime access for vehicles and workers.

Power

There is one thing theme parks need more of than anything else, and I'm not talking about guests.

Theme parks need a lot of power. In the case of Walt Disney World, 20 billion kWh each year.

That's a big number, but to most of us it's just a statistic. Let's think about it a minute.

If you turn on every light in your house, that might total one kilowatt.

So in a little over two million years you'd have equaled Walt Disney World's annual power bill.

Now if only we could find light bulbs that would last that long!

Because theme parks are so dependent upon power, the power distribution system must be designed with care. In June 2002, a small fire in an electrical substation closed Epcot. It was the first time in Walt Disney World history that an individual park had been forced to close for technical reasons. That's an amazing record of operational readiness.

Oh, and we also need to hide all that ugly power stuff from our guests. Remember, they came to escape from the real world. They don't want to be reminded they left the bedside lamp on.

Air Conditioning

One of the major consumers of all that power is air conditioning. People typically go to theme parks during the summer, when the weather is hot. A room full of hot, sweaty people isn't a pleasant environment. So theme parks work hard to cool them off.

In large theaters this is done by blowing cool air up underneath the seats and removing hot air near the ceiling. This carries the smells of all those hot, sweaty bodies up away from the audience.

When you consider the typical theme park has hundreds of thousands of square feet of air-conditioned space, it's easy to see efficiency is vital.

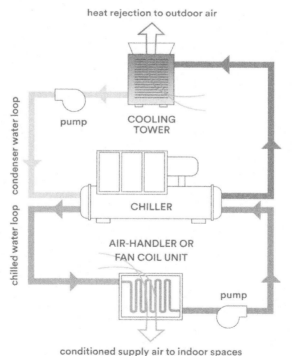

Drawing From Be-Exchange.Org

At large theme parks the most efficient way to air condition is by distributing chilled water from a central refrigeration plant to the individual attractions. The theory is that one gigantic heat exchanger is more efficient than many distributed ones, just as central air-conditioning in your home is more efficient than dozens of window air conditioners.

Of course, there are a few places that don't need air conditioning. We did a visitor center inside the arctic circle once. And there's always the first rows of seats in Shamu Stadium. That ice cold water will cool you right off.

Bathrooms, Sewer, & Water

Theme parks use a lot of water. During a typical day Walt Disney World reclaims 10 million gallons of the stuff, reusing it for irrigation or returning it to Florida's aquifer.

A lot of this water goes into bathrooms, so the efficiency of plumbing fixtures plays a major role in water conservation.

One of the more unusual jobs of my career was programming Epcot's Park Function Controller. This computer at Epcot Central was responsible for many park wide functions, including turning on the exterior lighting at different times each day depending upon when the sun set. Another role of the Park Function Controller was to control the automatic flushing of the urinals in the men's rooms.

Try putting that on your resume.

Picture by What's Up With the Mouse

Communications

When someone says "communications", most of us think about telephones. But theme parks have many different ways of distributing information.

Even in 1982 Epcot used fiber optics to transmit some data between Epcot Central and the various pavilions. In addition, the cable vault contained thousands of miles of copper wiring.

These days not just telephone but also computer data in many forms must be distributed everywhere within the park. Even cash registers, security systems and fire alarms need data lines.

A wise theme park designer will allow for new communications lines to be pulled through unused conduit, because it's a safe bet that the communications of the future will require interconnections that don't even exist today.

Photo by PDPics

More Site Work—Landscaping

Once the infrastructure—water lines, sewer lines, electricity, and communications conduit—is all in the ground, it's time to start pouring some concrete and building attractions.

...fade to calendar pages tearing away and fluttering off in the breeze...

Whew. That was fast. Those attractions look really great. But, ummm..., there's a problem.

This theme park looks like the Mojave Desert. Or worse. I don't think your guests are going to want to visit a place that has all the charm of a federal penitentiary.

That's why theme parks spend millions of dollars on landscaping. Walt Disney World alone has about 4000 acres of landscaping.

In some parks, the value of the landscaping exceeds that of the attractions. It's the finishing touch that creates a welcoming, relaxing environment for guests.

The selection of landscaping depends upon the theming of the park and the local climate. It was a real challenge building a jungle boat cruise in Tokyo Disneyland. The vegetation had to look tropical, yet withstand Tokyo's surprisingly frigid winters. Ever seen a hippo with snow on its back?

Tree Planting at Epcot's American Adventure Pavilion

Security

In a perfect world, guests would arrive after opening time, behave in an orderly fashion all day, and depart before dark.

Heh, heh.

In the real world they try to sneak in, drink too many beers, and hide in the bushes after closing time.

Photo by Ryan McGuire

Sometimes nature takes care of this problem. There was the guy who hid in the bushes next to Shamu Stadium at SeaWorld until the park closed, then went for a swim with the killer whales. No, they didn't eat him. He died of hypothermia in the chilled water.

But for problems that nature doesn't take care of, there's the security staff. They control access to the non-guest areas of the park, handle disturbances, prevent shoplifting, and make sure the park is clear of guests after closing. In times of heightened security concerns—and on grad nights—they also check for weapons and contraband. They're kept busy in other ways as well on grad nights, but we won't go there.

These days, for liability reasons, having a dedicated theme park security team and an integrated surveillance system is a must.

And ever since the 9/11 attacks, theme parks have continually raised the bar on security, with bag checks and metal detectors at the entrances to parks, scenes that would have been unimaginable in the past.

Now, with global pandemics a reality, we're in for wellness checks and certifications before guests will be allowed to enter. While that may not be 100% foolproof, much of the goal is to provide a sense of security.

Fire Control

There are many fire stations at Walt Disney World. Located near each of the major parks, they can respond quickly to emergencies.

Structure fires are rare, since all of the buildings have sprinkler systems. But there are also traffic accidents, and in the evening the occasional wildfire started by wayward fireworks.

Smaller theme parks may not have an official fire department, but they all have fire response procedures.

Emergency Medical

Walt Disney World also maintains a fleet of six ambulances, but, surprisingly, no hospital. There is a hospital not far from the property, though, and injured or sick guests can be transported rapidly.

Equally important are the first aid stations located throughout the parks. On a busy day a single theme park entertains more than 50,000 people. With that many visitors, injuries and illness are inevitable. So the wise park designer tries to provide guests with easy access to health care professionals.

On March 31st, 2002, the Wood-Mirlo family visited Disneyland. They were celebrating their son Dallas' eighth birthday. Mother Wendy was pregnant, but heck, the baby wasn't due until May 1st. So off they headed down Main Street.

Then Wendy's water broke.

The park staff came to life. Disney nurses raced to the scene, along with paramedics and the Disney police. They rushed Wendy into a security room and gathered around her.

"It's OK, mom," said Dallas, trying to comfort his upset mother.

Her husband held her head in his lap.

"I need to push," Wendy said.

The paramedic told her to hold on; he wanted to get her to a hospital.

It didn't work out that way.

Austin Ray Mirlo, 7 lbs. 5-1/4 ounces, was born at 11:30 AM.

"It's a boy!" the paramedic cried out. The crowd of Disneyland nurses, paramedics and police broke into applause.

The park invited Dallas to select two birthday gifts from a nearby shop, and the family was given return tickets for another day.

Austin was the third baby born at Disneyland.

Food Services

It takes a lot of people to produce 7 million inedible hamburgers a year. The logistics of staffing, stocking and maintaining a theme park's food service is equivalent to operating dozens—or even hundreds—of restaurants.

The best food service is themed to the park itself. Much of World Showcase's raison d' être—Tish, that's French!—is the variety of international cuisines served in its restaurants. Disney's Animal Kingdom offers a number of African themed dining experiences, and the Animal Kingdom Lodge serves authentic African cuisine and wines.

One has to be careful with this, sometimes. You won't find dolphin on the menu at SeaWorld.

Kitchens

Why are kitchens listed here, separate from restaurants?

Because in theme parks there isn't a one-to-one correspondence. Much of the food is prepared in central kitchens and delivered to the food service locations. Even sit down restaurants might be cleverly located so they open onto different areas of the park, yet share a single kitchen. Disneyland's Plaza Pavilion on Main Street and the Tahitian Terrace restaurant in Adventureland were really the same building.

The rationale for this is that kitchens are expensive. They have special facilities requirements that must be accommodated during construction. Not only are they major consumers of electricity, water and sewer, they also have to dispose of toxic waste: cooking grease.

Most restaurants have a grease handling system that routes grease to an underground tank outside, were it can be pumped out and disposed of appropriately. You don't want this job.

Merchandising

Let's not forget the final way that Disneyland upped per capita spending. Selling stuff.

Mouse ear hats and Tinker Bell wands cost next to nothing to make, and sell like hotcakes in a theme park environment.

Of course, not all theme park souvenirs are merely "stuff". At Universal Studios Florida and Disney's Hollywood Studios (in Orlando) you can buy authentic Hollywood memorabilia. And Epcot's World Showcase sells fine crystal and porcelain from around the world.

Accommodations

Here's the ultimate way to boost your theme park revenues: build hotels!

Be careful you don't build too few, like they did in the early days of Walt Disney World, when only the Contemporary and Polynesian resorts existed, and hundreds of low-end motels opened up in nearby Kissimmee.

But don't build too many, either, like they did at Disneyland Paris, where several had to be mothballed until attendance grew.

Disney pioneered the idea of continuing a park's theming right into the hotel when they partnered with the Disneyland Hotel, connecting it to the monorail ride. Fast-forward 35 years—The Animal Kingdom Lodge lets guests look out upon an African savannah filled with wildlife.

Commissioning

Once the park is built, we need to get it running. In fact this process starts while it is still under construction. Here are some of the major activities that take place as the attractions become ready.

Test and Adjust

Test and adjust. Those two little words make it sound so easy. See if it works, if not turn the knob. But there are tens of thousands of interrelated knobs any one of which has the potential to bring your attraction to its knees. And of course all that knob fiddling is done by exhausted individuals in an impossibly compressed time span.

So if it takes so long, why not schedule an appropriate amount of time to test and adjust from the start?

We did. The problem is that there are two certainties in any theme park project schedule:

- One is that the schedule will slip.
- The other is that opening day won't change.

Since test and adjust is the last activity before opening day... well, you get the idea. I've worked on projects where the scheduled time for test and adjust became negative!

Testing Electronic Cabinets At Escal'Atlantic, Aint-Nazaire, France

Test and adjust at its best is an exhilarating time. It's a time to get the computers that control the rides and shows programmed and working. It's a time to get speeds just right. A time to test the reliability of sensors and maybe move them an inch or two. A time to make sure that water doesn't overflow boat troughs and turn down the pumps if it does. A time to laugh when you discover the electrical conduits have been filled with water by those overflowing troughs

But what if the whole approach doesn't work? What if that system of a dozen high-pressure valves can't raise that lift in ten seconds no matter how you actuate them?

Then test and adjust turns into "insanely chaotic field redesign and punt".

Programming and Animation

While the engineers sweat out the details of programming the computers that run the ride and show, a completely different kind of programming is also going on.

Animated Figures and Control Cabinets, Epcot's American Adventure

Skilled—and specialized—artists work at an animation console, adjusting the way the Animatronic figures move. They run the show —or small sections of the show—over and over, tweaking and adjusting controls one by one to match the soundtrack and to make the movements as realistic as possible.

It's an arduous process, because for each figure they must learn its mechanical limitations and then take advantage of them. Sometimes a hip twist can provide the extra momentum needed to make that arm movement look just right. But then the figure is turned the wrong way for the next move.

Opening

This is what it's all about, and when it goes well, it can be incredibly rewarding. But it's not the end.

O C T O B E R

F S S M T W T F S S M T W T F S S M T W T F S S M T W T F S S

Epcot Calendar, 1982

Tweaking

There are always hundreds of details left, even after opening day.

Of course, the downside of the attraction working is that it's open to the public all day. This means finishing your tweaking in the middle of the night.

Adjusting Lift Sensors, American Adventure, Epcot

Renovation

Perhaps that sounds like the end of the story.

It's not.

Attractions typically have a life of seven years. Then the are reworked into an improved version, changed into something else, or completely dismantled. It's the natural life cycle of a themed attraction.

Demolition of Epcot's 17-Year-Old Horizons, 2000

PART 3
Professions

Creative People

Attractions start with the creative people.

Creative People fall into several camps. There are **Art Directors**, the guys who wander around the park holding up paint samples and fabric swatches. **Writers** come up with the basic story line of an attraction, and polish the script as it develops. **Set Designers** perform the same function that they do for live theater, designing sets, backgrounds, and props for attractions. **Media Designers** produce the audio and video materials that will be played —over and over and over—in each attraction. Their job is similar to that of filmmakers, although the latter have larger budgets. And there are hundreds of other professions that may also be involved in the conceptual phase of a new attraction: composers, actors, story board artists, and so on.

Once they figure out—more or less—what they want to do, that's when the technical people become involved. They figure out how much the whole thing will cost and then tell the creative people they can't afford it.

The creative people go off in a huff, change a bunch of stuff, then present their new ideas. About half the time the new way actually is cheaper than their first approach.

Anyway, you get the idea. There's a lot of give and take before the attraction's initial design is baselined, and even more as the details are firmed up, usually while we're trying to build it. Changes can make for an interesting time, all right.

Directors and Producers

While we were building Epcot we would often see its two main Art Directors, John Hench and Marty Sklar, walking around the partially constructed park. John would invariably be holding a handful of color swatches and Marty would be talking excitedly about the park. All of John's color swatches seemed to be pink or purple, and I sometimes wondered if anyone but him could really see the difference between the dozens of different shades. But he knew what he wanted, and he got it.

In addition to John and Marty, there were also Art Directors for each Epcot pavilion. Some of them, like Ward Kimball who did World of Motion, had been friends of Walt Disney's, and had made Disney movies before they got involved with theme parks. Others had grown up with theme parks.

Although they took completely different approaches, and used different story telling and visual techniques to tell their stories, each had a clear concept of how they wanted the attraction to turn out, and they worked throughout the project to keep the end result as close to that concept as possible.

In film production, the Art Director is the person who is responsible for the selection of colors, fabric, and is responsible for the overall look of the movie. In theme parks their role may be far greater. Because of this, there is an increasing trend to call these people **Show Producers**. Sometimes the Show Producer will have an Art Director working for him or her, other times they will fill that role themselves. If the attraction incorporates a film, the Show Producer is often the Film Director, too. The Show Producer is ultimately responsible for the content of the entire attraction.

One of the best Art Directors in the business is Bob Rogers. Bob is an Academy Award nominated director who worked with Disney on Epcot, Expo '86, and Space Center Houston. Bob's company, BRC Imagination Arts, has a long-standing relationship with General Motors. Bob designed both generations of postshow theaters at Epcot's General Motors pavilion, and designed the brilliant Mystery Lodge show for the GM sponsored pavilion at Expo '86.

Bob's most memorable attraction from Epcot days was The Bird And The Robot. In this lighthearted introduction to automation, an audio animatronic parrot—smoking a cigar—bosses around a cooperative robotic arm. The contrast between the artistry of the animatronic and the industry of the robot fascinated guests, and did a highly effective job of presenting General Motors' corporate message about the sophistication of their assembly lines. Bob credits Marty Sklar with the original idea for this attraction. Whoever thought of it, the implementation was brilliant. It was one of the most popular shows at Epcot before succumbing to a burst of spring-cleaning in the late '90s.

Although Bob's title on The Bird And The Robot show was probably Art Director, his role was really more like a Show Producer. Good Show Producers are hard to find. They must balance the needs of all the different people working on a project—both creative and technical—while keeping track of money and schedule. Show Producers spend a lot of time arbitrating disputes. They also become easy scapegoats when things go wrong. And they aren't always aware of problems in the making if the team members aren't honest with them. It's a tough job.

Writers

In 1982 construction began on a theme park between Walt Disney World and the neighboring town of Kissimmee. It was to be called Little England and would feature a medieval village where guests would participate in a Renaissance fair, complete with bawdy revelry, buxom serving girls proffering tankards of beer, and lots of eating with your fingers. After only a few months construction stopped. Today it's part of a housing development.

What went wrong?

It was the classic problem of story and audience. What was the story supposed to be? Why were we sweltering in the summer heat of Central Florida while pretending to be in a medieval village? And who was supposed to come to this theme park? A sizable percentage of the tourists who visit Walt Disney World are from England. Did anyone really think that these visitors would be interested in flying to Florida so that they could pay to get into a recreation of... England?! And nearly all visitors to Central Florida are families. Would they be interested in a "bawdy" experience?

Obviously, the answer was no.

That's where writers come in. If our writers do their job there'll be a reason for people to visit our medieval attraction. And the activities there will be appropriate to our audience.

When I think of writers I imagine an unshaven recluse, working in a lonely garret, cigarette dangling from one lip as he pounds out page after page of torrid text. Theme park writers aren't like that. They spend most of their time working with other members of the creative team. Most of their time is spent developing and refining ideas with these other people, in meetings and informal discussion groups.

But there is one way that theme park writers are just like my imaginary writer. They emphasize "backstory". Backstory is a term writers use to describe what happened before the beginning of the story. Backstory increases the believability of fiction because it guarantees each thing that happens in a novel has an underlying reason, even if we don't see it first-hand. For the same reason, backstory greatly increases the believability of a theme park attraction. If every themed item we encounter has an explanation for its existence—rather than just being empty set decoration—it will lend an aura of credibility to the attraction that the guests will perceive.

A terrific example of this is defunct Adventurers' Club at Pleasure Island in Walt Disney World. The Art Director spent months traveling around the U.S. visiting flea markets and swap meets, collecting oddities from the past. The walls and ceiling of the entire attraction were peppered with these items.

But this was no neighborhood restaurant with knick knacks nailed to the walls. The writers worked out the backstory for hundreds of these objects. Small signs on many of them identified the (fictional) club member who procured the item, along with the time, place and circumstance of its acquisition.

They also created a club motto, and even a club song that visitors had to learn. Costumed performers portraying club members wandered through the space, interacting with visitors. They weren't just reciting memorized acts. They were constantly ad-libbing while remaining in character. For these roles to work, the performer had to be familiar with the entire backstory of his character. Portions of this backstory were played out, but others were never explicitly stated. Yet their existence lent credence to the whole.

Backstory can also backfire. The entire Pleasure Island complex is a good example. Each building had its own elaborate backstory—something about Merriweather Pleasure creating the place, and Fireworks blowing up one building, and they managed to work in a roller skating rink and a disco. Even reading the sheaf of explanatory papers they handed out to Operations on opening day was mind-boggling. The little brass plaques in front of each building were incomprehensible. How to convey the story to uninitiated tourists?

The answer is, they couldn't. It wasn't particularly well attended, and some of the venues were downright unpleasant. Operations took ownership, and over the course of a year added "streetmosphere"—stuff happening outside to give the place some character—in the form of redemption games, a live stage, and a DJ. Inside they re-themed several of the clubs and restaurants, converting a mall-like food court to a jazz club and the roller rink to a beach resort. (I hope the flooring contractor wasn't around the day they dumped sand all over his hardwood roller rink.)

Today, Pleasure Island is gone. A conscious decision was made to abandon it entirely.

Artists

Scenic Designers are nothing new. For over a century they have worked in films, creating the sets that are the backdrop of the production. The history in live theater is far longer. The role they fill in theme parks is much the same. There are important differences though.

Theme park sets are larger than many movie sets. More significantly, they are all-encompassing. In a movie the camera points in only one direction. On a ride our eyes are constantly in motion left, right, up and down.

Like their counterparts in live theater, theme park sets must meet fire and building codes. This applies to both the selection of materials, and the way the materials are used. Often the sets are in an area of the ride that is also used as an evacuation area. This must be taken into account by the Set Designer.

Theme park sets must be durable. A Broadway show may last only months, with only a few actors and stage crew ever near the sets. But hundreds of maintenance people over the course of seven or more years will interact with theme park sets.

In a stage show or a film all of the technical junk is offstage. But in a theme park, the sets must conceal speakers, lighting fixtures and sometimes control equipment.

Occasionally stage show sets use forced perspective. The technique is one in which objects that are farther away are rendered much smaller to increase the sensation of vast distances. It is used less often in films. But nearly every theme park ride set incorporates forced perspective to expand the space far beyond the physical boundaries of the scene. It's a complex technique to render believably, and is made more difficult by the fact that the set is viewed from many different angles. Objects that appear far away as you enter a scene must still seem far away as you exit on the opposite side, even though you have moved past them.

Stage shows often use flat sets. But if you're moving past them that doesn't always work in a theme park. And even the best seats in the house are still far away from the sets in a stage show. A little distance can cover a lot of sins. Anyone who has ever seen the sets from a motion picture or television show will know how crude they can be; the camera lens is very forgiving. That's not true in a theme park, where it's not uncommon for the ride vehicle to pass within three or four feet of the sets. That means that rocks need to look like rocks, wood like wood, and metal like metal. Yet no one is going to fill a theme park ride with real rocks, or spend millions of dollars on complicated metalwork. Instead inexpensive wood, fiberglass and other materials are combined, then painted by artists to look—in some cases, anyway—better than the real thing.

Only occasionally do sets in stage shows serve a structural purpose. But in theme parks it's not uncommon for them to provide access to higher levels, so that maintenance personnel can reach lighting fixtures, or reach elevated equipment areas. They're also used to support special effects projectors, fans, and other devices.

Sometimes theme park sets incorporate moving pieces: a giant printing press; window shutters that blow in the wind; or spinning wheels on upended carts.

Theme park sets also use a wider variety of mixed media materials than theater sets. Ultraviolet paint is common—sometimes sets need to completely change appearance when lighting changes from normal to ultraviolet. Theme park sets also incorporate more landscaping: dirt, bushes, trees, even flowing mud.

When they constructed the sets for the queue area of Disneyland's Big Thunder Mountain Railroad, one employee's full-time job was to "age" wood. He spent weeks whacking perfectly good wooden beams with an axe to create the look of wood that had been exposed to the elements for many years. Some cleverly applied paint to add burn marks and water stains, and you've got an instantly decrepit railway.

Most people don't spend a lot of time thinking about lighting. You walk into a room, flip on a switch, and don't give it a second thought. But lighting in a theme park can make a thousand dollar set look like a million bucks.

Halfway through the American Adventure stage show at Epcot, the house is plunged into darkness. We hear the crackle of a fire and the sound of crickets in the wilderness. At center stage a campfire begins to flicker. Standing before it is Chief Joseph, leader of the Nez Perce. He flings his arms wide, and the light of the fire catches the colored stripes of the Indian blanket wrapped about his shoulders. Firelight plays across the crevices of his face as he delivers his famous words, "From where the sun now stands I will fight no more, forever."

What would this scene be without lighting?

Well er, dark. But anyway, you see what I mean. In this case a talented lighting designer took a good scene and made it great.

Far more than in a stage show, lighting in a theme park is up close and personal. It's used to great dramatic effect, to increase the realism of set pieces, and also to reveal different physical elements as the story unfolds, a technique rarely used in live theater or films.

Media Designers

Your vehicle rounds the corner as an explosion rocks it. A brick wall disintegrates. Bricks bombard your vehicle, each impact producing a rapid-fire crack, crack, crack! Twisting around, you surge high above the New York streets, but another impact swings you around, and you are suddenly plummeting faster and faster toward the ground. At the last moment some web-like thing arrests your forward progress and you dangle, nose down, staring at the pavement.

What's going on here?

It's the climax of the Spiderman ride at Universal's Islands of Adventure theme park. From a media production standpoint, this is the world's most complex amusement park ride.

The ultra-realistic experience is created by the perfect synchronization of vehicle motion to projectors and onboard surround sound. To get this all to work together is a control system nightmare. But to produce the media was an act of brilliance. Not only does the 3-D movie have to match the surrounding sets, it has to match the motion of the vehicle through the sets, since perspective changes from one side of the scene to the other. And the sounds that reinforce the action need to be consistent with the acoustics of the imaginary city streets through which you're flying.

Many, many man-years were spent developing the CGI film and accompanying audio track. And the final product shows it. It's the pinnacle of theme park media design.

Video can also be used for special effects projection. These are used to create backgrounds of waving wheat, wildfires burning against a sunset sky, rain falling, or clouds floating past.

In addition to visual media, Media Designers are also responsible for music and other sounds throughout the attraction. They don't actually compose the music, but they do produce it. For most themed attractions the music must match action, scene length, or vehicle movement. This means—usually—short playback lengths and easily recognized themes.

Most themed attractions also use sound effects and ambience. These are nonmusical sounds that reinforce the experience. Sound effects usually synchronize with observable motions: the clatter of a moving chain, explosions, a hammer pounding a nail. Ambience is background sound that helps set the mood: water lapping against a boat, seagulls crying overhead, the sounds of children in a schoolyard.

The Media Designer's job is to acquire, edit, process, and encode the material, ready for the Audio/Video Engineer's use. Unfortunately this process usually occurs late in the development cycle, and the Audio/Video Engineer often must struggle with last-minute installation of untested media, and hope for the best.

There's one other type of media production that is common: software. Interactives, information terminals, and many exhibits use computer software as audio and visual complements. In this case the responsibility for the development of the end product falls not with the programmers who implement it, but with Media Designers who are skilled in the creation of audio and visual software.

This is a highly specialized field, because it combines both technical and artistic skills. Good media designers are highly paid; they can make millions developing games for home video consoles or cell phones. So it's sometimes difficult to attract top talent for one-of-a-kind theme park projects. Nevertheless some truly stunning results have been produced.

Technical People

The Technical People are the ones who bring the Creative People's dreams into the world of reality. From the architects to the engineers to the technicians, they are the ones who figure out how to do it and then get it done.

Meanwhile, all those changes the Creative People make are driving the Technical People crazy!

Mechanical Engineers

Anyone who likes to spend an evening disassembling their automobile's engine is going to love spending the day designing a ride vehicle.

Mechanical engineers who work in theme parks work with slightly different constraints than most mechanical engineers. In the real world, mechanical products are generally classified as high-volume or low-volume.

An example of a high volume product is the transmission of your car; there are millions. High-volume products tend to be carefully designed, sparing no expense, to guarantee that each will cost as little as possible to produce. Given the competitive nature of the automobile industry, every dollar counts, particularly when you multiply it by millions of units.

A low volume product is something like the crawler that moves the space shuttle from the vehicle assembly building to the launch pad; there are only a few. Low-volume and one-of-a-kind items are generally built without many cost constraints, but may not be intended for repetitive wear and tear applications. That crawler may only make a few hundred trips during a twenty-year lifespan.

Theme park attractions are different. They require mechanical designs that are produced in relatively low volume at moderate cost that will be extraordinarily durable and reliable. Unlike high-volume products, anything needed for a theme park must be developed quickly, without years of research and testing before deployment. And unlike the shuttle crawler, theme park mechanical designs are used several times a minute.

Here's an example. The brake mechanism on a roller coaster must be extremely reliable. Yet that vehicle goes around the track in under three minutes, loads in a minute, then does it again, all day long. But the guy who designed it had a staff of three and less than 18 months from conception to deployment. That requires some careful and competent engineering.

The tight time frames and small staffs involved in theme park design might lead you to suspect that theme park designs are not all that complicated. That is far from the case.

Here's another example. Rides like Disney's Haunted Mansion use a propulsion mechanism called an Omnimover. It consists of a continuous, motor-driven chain that drags the vehicles around a ride track with almost any horizontal or vertical profile. Mechanisms on the underside of the vehicles cause them to rotate into desired orientations at various spots around the ride. There may also be mechanisms that trigger audio or cause mechanical actions such as lowering the lap bars.

Rides like this contain thousands of moving parts. Yet the system was designed by a small staff of mechanical engineers in a short period of time. It has proven highly reliable in over 40 years of nearly continuous use.

While the majority of mechanical engineering in theme parks involves wheels and vehicles, theme park Mechanical Engineers also work with water pumps, electrical winches, and hydraulic lifts, plus anything else it might take to satisfy those crazy Art Directors.

Earthquake Flood Valve, Universal Studios Florida

Architectural Engineers

Just as Art Directors come up with ideas that challenge the Mechanical Engineers, they also challenge the Architectural Engineers. Theme park buildings may have huge cantilevered spaces, or be made from unusual materials, or not even look like buildings.

But then again, so may the neighborhood shopping mall. So you might think architectural engineering for theme parks is a lot like architectural engineering elsewhere. But shopping malls are rarely built on reclaimed land, rarely have to withstand the constant vibration induced by careening ride vehicles, and almost never need a roof structure strong enough to support catwalks laden with lighting and special effects equipment.

There's a difference between an architect and an architectural engineer. An architect is responsible for designing a building that suits the owner's needs; an architectural engineer makes sure that building remains standing.

As with theme park Mechanical Engineers, theme park Architectural Engineers must work on tight time frames, yet create innovative spaces that meet both the Art Director's expectations and the local building codes. They must work carefully and competently to achieve exciting and yet safe results.

Glenn Birket Atop America Adventure, Epcot, 1982
(OK, He's Not an Architectural Engineer,
but Those Are Plans!)

Structural and Civil Engineers

What about all that big stuff that isn't a building? Roads, bridges, lagoons, even parking lots are the domain of Structural and Civil Engineers. That stuff may not seem as exciting as some of the other elements of theme park design, but it's just as important.

When Walt Disney World was developed, the first step in taming 44,000 acres of swamp was creating a canal system that would allow the land to drain. And the second step was creating roads so the construction site could be accessed. That work continues to this day.

Civil Engineers at theme parks perform a parallel role to that in any big city. They do site planning, transportation studies, and energy distribution. They're also responsible for environmental issues including wetlands preservation, water use, and waste disposal.

Sometimes the requirements are just like those of a small city. Other times they can be far different. Employees commute to a theme park. But so do 40,000 or 50,000 visitors, who may arrive by car in just a couple of hours. And unlike commuters, they don't know where they're going! So theme park traffic signs need to be a lot better than those on the interstate highway.

When Disney decided to build Walt Disney World, one of the first things they did was to get the Florida legislature to make their vast property into its own pseudo-county, the Reedy Creek Development District. This gave them a rare opportunity to try new ways of doing things with a fresh building code. For example, underneath the Magic Kingdom they created a network of "utilidors" for the distribution of food, services, and as a way for employees to easily move about "backstage".

Along the ceiling of these corridors runs plumbing, electrical circuits, control wiring, and the innovative "AVACS" system. Basically, it's a giant vacuum cleaner. At various spots around the Magic Kingdom there are trashcan-like receptacles that employees use for waste disposal. But they're connected to a quarter mile of tubes that lead to the energy plant behind the Magic Kingdom. Close the lid on the receptacle and your trash is sucked out of the Happiest Place on Earth and into the trashiest place on earth.

AVACS was a pretty ambitious and revolutionary bit of civil engineering. To really work, the system had to be powerful—so powerful that they tested it using automotive batteries. And a system that powerful has to be safe. Interlocks prevent the vacuum from engaging if any lid is open. The necessity for this was demonstrated clearly one day. A hatch was left open on receptacle inside a small shed. When the vacuum came on the entire shed turned into a wad of crumpled aluminum! (My vote for worst job in the world: the little guy who has to slide on a cart through the tubes, cleaning out blockages.)

The other revolutionary aspect of AVACS was what was at the other end: an energy plant where trash was turned into energy to help power the park. You'll note my use of the past tense. Unfortunately, like many innovative systems, the energy plant cost more than it saved, and it was abandoned in the 1980s for much cheaper electricity purchased from outside sources. But you've got to admire them for trying.

Ride Control Engineers

It takes nerves of steel to be a Ride Control Engineer. Also a sharp intellect, and good analytical skills, because the safety of— literally—millions of people is dependent upon the Ride Control Engineer.

This safety is achieved in a number of ways. The most common technique is redundancy: more than one computer making decisions about the ride. If the computers disagree, the ride E-Stops—industry parlance for an emergency stop, a controlled event that brings things to a motionless safe state as quickly as possible.

Most Ride Control Engineers are expert at failure modes and effects analysis. Using this technique, every single monitoring or control point in a ride is analyzed to see what the result would be if it failed. Single and multiple point failure analysis and formal safety acceptance testing are also essential.

Finally, if the ride is to be a good one, the Ride Control Engineer must assure that it has the motion profile the Art Directors intended, and that it is properly synchronized to other systems.

As you can see, Ride Control Engineering is one of the most demanding yet rewarding careers in theme park design.

Ride Atop the 1/4 Mile Tall Stratosphere Tower, Las Vegas, Photo by Michael

Show Control Engineers

With more variety and less stress than ride control, Show Control Engineering gets my vote as the most interesting job. Where else can you find a job where you get to control dancing alligators, bubbling lava, and singing broccoli—possibly all in the same day?

The show control engineer has a hand in nearly everything you experience from the moment you enter an attraction.

Let's put on our Show Control Engineering hard hat and walk through a simple theater attraction. Here's how the Show Control Engineer views it:

As we enter the preshow doors, we pass through an electronic turnstile were an infrared beam counts us and displays the number of people in the preshow area on a small display near the Operator Control Console (OCC). Over there on the wall, designed to fit into the themed paneling, the countdown clocks tells us there are 18 minutes until the next main show. The background music in here has been running continuously since it was started by the show control system early in the morning, providing a general ambiance. That was also when the show control system set the lighting levels.

Ten minutes before the main show begins, the show control system fades out the background music and starts a video that runs on monitors in the preshow area. The Show Control Engineer makes a note to bring the lighting levels down a bit more to make the video easier to see. Periodically, the hostess needs to ask people to move all the way to the front to make room for more. When she does, she presses a button on the microphone. This causes the show control system to automatically "duck" the soundtrack and mix in the microphone audio.

As main showtime nears, the show control system fades up a spotlight to illuminate the preshow hostess and patches the microphone audio through to the overhead speakers. The hostess introduces the show. Partway through her spiel a cue light on the OCC warns her that the main show is about to end. She cautions the guests to remain behind the yellow line until the doors open. A moment later the show control system flashes the "Caution: Automatic Doors" sign above the entryway. After a few seconds' delay it commands the door controller to open the doors.

The audience begins to file through the entry doors even as the previous audience is leaving through the exit doors on the other side of the theater. The show control system has set the theater houselights to bright so people can see clearly. It has also started the fill/spill music in the theater which is a perfectly synced continuation of the music in the preshow. The show control engineer is pleased.

After a minute or two the show control system commands the theater exit doors to close, and in another minute the theater entrance doors also close. OCCs near both sets of doors allow the operators to override these automatic actions if needed.

Once everyone is seated, the hostess at the front of the theater presses a button on an OCC to tell the show control system to start the main show. As she gives her introductory spiel, the show control system starts up the projectors and audio source and ensures that they remain locked together in perfect synchronization throughout the entire 20 minute show.

The houselights gradually fade to darkness. The audience waits in breathless anticipation—the show control engineer crosses her fingers. As the hostess completes her spiel the curtains are commanded to open. Once the screen is exposed, the show control system starts the movie. For the audience the main entertainment is just beginning. But for the show control engineer the action is pretty much over.

Sometime during the movie, out in the preshow the cycle begins again.

At the completion of the movie the process is reversed: the house lights come up, the curtains close, the exit doors open, and the guests Exit Through Retail.

And that was a simple show! Imagine all the buttons, lights, wires and logic it took to do that, plus handle unusual conditions such as building-wide pages, fire alarms, failed projectors, and curtains that overshot their mark.

More than any other single individual, the show control engineer understands how the entire attraction operates.

Linda Alcorn, WDI EuroDisney Field Office, 1991

Audio/Video Engineers

With the Show Control Engineer responsible for controlling all that audio and video, you might be wondering what the Audio/Video Engineers do.

They source it, process it, amplify it, and deliver it.

Theme park audio generally doesn't come from tape recorders or CD players. The devices used to playback audio in a theme park must work for years without any maintenance—not even dusting. So mostly they are solid-state. The best units use the same kind of storage card that you might plug into your digital camera.

Similarly, video must also require no maintenance for years. Most theme park video players use magnetic disks or solid state media to store high performance digital video.

As we've seen from our description of Show Control Engineers, there are many different audio zones with different control requirements: looping, synchronization to movies, microphones, etc. The Audio/Video Engineer connects all of these sources to some sort of digital processing equipment that can be used to alter levels and frequency response, and mix and route the audio to different areas.

The output of this audio processor is then connected to the amplifiers. From there it is distributed around the building. The types of amplifiers and distribution vary depending upon use.

Video is sourced and distributed similarly. Video monitors, flat panel displays and digital projectors must be selected for appropriate size and brightness.

The audio video engineer also specifies all of the speakers. There are many different types of speakers in a single attraction.

Overhead ceiling speakers are common in the preshow. A full range of speakers—from subwoofers to horns—are used in the main show. And outdoors there may even be speakers disguised as rocks!

To select the amplifiers and speakers the Audio/Video Engineer needs to understand something about the acoustics of the spaces in the building. This is not to say that Audio/Video Engineers are acousticians. This function is usually performed by a specialist, or by the architectural engineering firm. They are the ones who select the acoustical treatments in the building. (Project management then deletes most of them because they are deemed frivolous and expensive and "you can't see them anyhow".) The Audio/Video Engineer must deal with the result.

24-Track Tape Binloop Machines Used for Audio at Epcot in 1982

Audio/Video Engineering is a great career choice for someone who spends his or her weekends perfecting the sound of their expensive stereo. In the early days of theme parks, audio systems and acoustics were afterthoughts. In the past 20 years a demanding public has come to expect superior audio performance. That means lots of expensive toys for the Audio/Video Engineer in spite of the project manager.

Lighting Designers

Nearly every stage show has a lighting designer, and it might seem that lighting design for theme parks would be little different than ordinary stage lighting. This may be true in the case of live shows, but most theme park attractions are static and highly automated. This means that exactly the correct lighting fixture can be selected for each feature that is to be illuminated. And theme park Lighting Designers aren't just "throwing" light. Often the fixtures themselves become a part of the theming.

Theme park attractions use a wide variety of esoteric lighting fixtures: spotlights, architectural lighting, spark tubes, liquid neon, strobes, lasers, and fiber optics are all common.

In a live theater, a lighting fixture may only be on for a few minutes a day. In a theme park it will be on for years. This means that careful selection of lamp type and color filter are essential in order to keep down maintenance costs, and keep the same "look" year after year.

Systems Engineers

While the Show Control Engineer may understand how most of the attraction works in microscopic detail, the Systems Engineer is the guy responsible for making sure it works at some higher level.

If you like sitting at a lab bench doing experiments, or designing things on paper and then seeing them "in the flesh", then you'd hate being a Systems Engineer. On the other hand, if you like to look at the big picture and hate getting caught up in annoying little details, then Systems Engineering is a perfect career choice.

A lot of what Systems Engineers do is similar to what Coordinators do: they make sure people communicate. But unlike Coordinators, Systems Engineers operate on a technical level. They understand what the Mechanical, Electrical, and Electronic Engineers are trying to accomplish and they make sure those disciplines understand each other's needs.

It can save a lot of money if—when everyone gets to the field—things work together as intended.

I've often seen attractions where the behavior of the mechanical parts—which seemed so obvious to the Mechanical Engineer—is completely different than what the control systems engineers expected. As a result, weeks of time are wasted trying to re-architect the software or add new electronics to deal with a behavior that the Mechanical Engineer knew about all along.

A good Systems Engineer prevents this from happening.

The Systems Engineer evaluates trade-offs, determining whether more money should be spent on the mechanical system to make the control systems simpler and cheaper, or if more money should be spent on the control system allowing for a less expensive mechanical design.

Systems Engineering is a relatively new approach to theme park attractions design. Sometimes it works, sometimes not. I've seen Systems Engineers accomplish next to nothing over the course of a three-year project. But I've also seen them save the project millions of dollars by anticipating problems before they occurred and solving them while it was still inexpensive.

Steve at Epcot, 1982

Project Engineers

Project Engineers are theoretically in charge of all engineering on a theme park project. This means everything from buildings standing up to the ride working to exit signs lighting and even toilets flushing. It's all under their jurisdiction.

This may sound like a terrific responsibility, and it is.

In practice, Project Engineers tend to focus on where the money gets spent. That means the building. As a result most Project Engineers are "steel and concrete guys". This means that they know little about control systems, they rarely become involved in audio or video, and don't regard special effects as an engineering discipline at all.

Because they spend most of their time working with the architects and construction companies, those of us who are focused on making the attraction work seldom have much to do with the Project Engineer. But it's an important role, and one which has the potential of saving the company more money than the entire cost of the control system or audio installation.

It's easy to get Project Engineering confused with Project Management, and even its relationship to Systems Engineering can be fuzzy. In a nutshell, the differences are this:

The Systems Engineer is responsible for making sure that the mechanical, electrical, and electronic systems function together. His concern is with the ride and show content of the attraction. It does not involve structural engineering.

The Project Engineer is responsible for all engineering activities, but focuses on structural issues.

The Project Manager is responsible for the overall project, but he or she does not have an engineering degree. It's the Project Manager who makes sure the checks get signed.

So who is the boss?

A lot of people working together make a theme park attraction happen.

As engineers our ultimate boss is the Project Engineer, who paradoxically has little interest in most of what we call Theme Park Design.

Collectively, the big boss is the Project Manager. He or she is responsible for all aspects of the attraction, and reports directly to Corporate Management.

But in the end the boss is... You.

As a Theme Park Engineer, you are ultimately responsible for your own systems. In the mad rush that always occurs to complete an attraction, no one can second-guess you.

You have to make the right decisions or the attraction won't open on time.

Putting the Team Together

Architects design the building itself, and Architectural Engineers make sure it will stand up. Given the fanciful structures of many themed attractions, this is no small feat. They are generally Structural or Civil Engineers.

Mechanical Engineers design nearly everything that moves: vehicles, set pieces, props, ride doors. They also work with hydraulics (oil or water) and pneumatics (air). Some smaller mechanical items may be handled by animators or special effects artists.

Systems Engineers unify all of these other disciplines. They often have oversight of the entire attraction's engineering, but are seldom intimately familiar with any single subsystem. The Systems Engineer is often the one who is ultimately responsible for ADA (Americans with Disabilities Act) compliance and meeting applicable safety codes.

Technical Writers document what we did. They write maintenance and operation manuals. A common problem with Technical Writers is that they may not fully understand what they are writing about. For this reason, there is a trend toward the design engineers preparing their own technical documentation.

Lighting Designers fall somewhere between the Creative People and the Technical People. The selection and location of fixtures and colored filters is certainly an artistic endeavor. But the selection of dimmer cabinets and control equipment is a technical one. Sometimes the show control engineer participates in the technical portion of this task. And we're always grateful when the Lighting Designer has an idea of the technical ramifications of his or her design.

Special Effects Designers also fall somewhere between the Creative People and the Technical People. They have long, scraggly beards, and walk around hunched over, like forest gnomes, muttering about smoke fluid and synthetic mud. Special Effects people often design their own electronic control boxes to activate their effects. But not if the show control engineer can help it. These two disciplines have a long history of miscommunication. I think it has something to do with the smoke fluid.

Alcorn McBride Exhibit at IAAPA, 2018

Planning, Scheduling, and Finance

Coordinators are like honey bees. They flit from discipline to discipline, collecting the needs of one and passing it on to the next. As opening day gets closer they flit faster and faster.

Planners and Schedulers use project timeline software and spreadsheets to put together highly detailed schedules that are read only by upper level management. The rest of us are just working as fast as we can and making up bogus percentages of completion to keep them happy.

From the very beginning of the project the Planner tracks every discipline, following their needs and establishing the linkages between each task and the deliverable that preceded it. Gantt charts keep track of everything.

The Gantt chart is a horizontal time line that could be days, or weeks, or even hours, depending upon the need. Parallel lines for each task are drawn between starting and ending points, with milestones shown as triangles.

Notes indicated which tasks need to be completed before others could start. If one task stretched, it could force dozens of others to slide further down the timeline. Sometimes the chart could also be used to figure man loading, which helped the finance people calculate budget.

Estimators and Financial Analysts are the people who figure out how much more the attraction is costing than it was supposed to, and then try to explain why.

Project Managers are the guys who think they're running the project, when really it is a three-ton bull escaped from its pen and bearing down on opening day no matter what happens. Along the way they organize lots of meetings where they try to encourage communication between the various engineering disciplines, construction personnel and Creative People. One of their vital functions is to bring lots of pizza in the middle of the night. In the end, if the show doesn't stink, they get the credit.

Epcot Calendar, 1982

Operations People

It takes people to run a theme park. Something like 100,000 work at Walt Disney World alone. When we visit a park we see the costumed characters, the ticket takers, the kitchen workers, and maybe a few custodial people. But there are hundreds of people behind the scenes for every one we meet "on stage".

Who are all these people, and what do they do?

Ride Operators and Other Cast Members

Operations personnel are the most visible people in the park. These are the people who interact with the guests the most, because they operate the rides. That can mean anything from taking tickets to driving a boat while delivering a comedy monologue on the Jungle Cruise.

Many operations people are "casual temporaries" which in theme park parlance means "don't call us, we'll call you". But an equal number are full time employees, who after some experience become "leads" of attractions or perhaps whole lands. They in turn report to supervisors responsible for operation of the entire park.

Maintenance

There's a lot of stuff to break in a theme park. Everything that moves, everything that has electricity or water—but hopefully not both—flowing through it is liable to break at some point or other. It can take hundreds of maintenance people to properly maintain a large theme park. Some of these jobs are quite specialized: projector maintenance, vehicle mechanic, computer control systems specialist, audio technician, even scuba divers to keep aquarium facilities maintained.

In some parks the custodial personnel also report to Maintenance, but in others it is a separate department.

Management

These are the guys who run everything. At the level below president will be vice-presidents responsible for the aforementioned creative, technical, and operational departments, plus a number of overhead departments including Finance, Planning, and Marketing.

One of my friends runs a department devoted to figuring out how many guests will be in each park at a major Florida resort on every day for the next five years.

Think about that.

What factors could influence it? Day of week, school holidays, weather patterns (including the likelihood of tropical storms during a given week), the Super Bowl and other television or world events, not to mention whether we theme park engineers are designing them a new attraction for that season.

And why does anyone need to know this?

What if Mondays in winter have an average attendance of 15,000 people, and then 40,000 people show up at your park one Monday because it's a school holiday? You're gonna need a heap of burgers to satisfy that crowd. Plus more ride operators, food service personnel, cashiers, custodians, and on and on. Time to call the casual temporaries!

Clearly a theme park is a complex collection of inter-related disciplines, and there is a lot more going on behind the scenes than the average visitor would guess.

PART 4
Technical Stuff

Technical Stuff

In the first edition of this book I included quite a bit of technical information in the professional descriptions of the electronics engineering disciplines of show, ride, audio and video design.

- Some readers thought it was too heavy on engineering.
- Some readers wanted more engineering.
- How to solve the problem?

In this edition I decided to separate the engineering stuff into its own section, and supplement it. If you're interested, get ready for a deep dive into the details. And if not, just skip to the Epilogue!

Show Control

The appetizers and salad were fun, but now we've got some real work to do. Now we'll put on our Show Control Engineer's (hard) hat and get down to the real work of designing the controls for a theme park attraction.

Show Control Cabinets at American Adventure, Epcot

Types of Show Control

Perhaps you already have some preconception of what show control systems look like. Maybe a roomful of supercomputers with cables dangling everywhere. Or banks of hamsters in wheels with little bicycle chains running this way and that. Or how about a shaft lined with revolving discs about the size of 78 RPM records, each one carved into unique shapes to control the motions of animated figures?

Believe it or not, one of those actually was the show control system for the original Pirates of the Caribbean attraction at Disneyland. Can you guess which one?

If you guessed the 78 RPM records, you're right. Shaved into different shapes, they controlled the movements of the animatronics.

Today's show control systems are a bit more complicated than that. The form they take depends upon the function they must perform. But before we describe what they are, let's look at what they aren't.

The Case Against PCs

In general, show control systems are not personal computers. There are several good reasons for this.

At first it seems like a great idea to use PCs for show control. They're easy to get, everyone knows how they work, and they're so cheap these days they're almost a commodity.

But the design life of a theme park attraction is at least seven years. Many are expected to operate for 10 or 20 years. During this entire time they must be maintained. In order to maintain them you must be able to obtain spare parts. Suppose lightning damages part of the show control system. In the middle of the night someone has to diagnose it, find replacement boards, install one, and verify the system works exactly the same way it did before the lightning strike.

Have you ever tried to replace a component in a PC that was several years old? The circuit boards used in today's PCs won't even plug into PCs made more than a few years ago! Besides, a modern PC's motherboard incorporates most of the functions that used to be individual cards. What if the original system simply won't function the same way on newer, faster, different hardware?

Since PCs are certain to change dramatically from year to year, the only solution to this problem is to stock multiple spares at the outset. But this eliminates any price advantage that the PC might have had.

The fact that everyone knows how PCs work also proves to be a disadvantage. Because the maintenance person knows how to configure his home PC, it's a great temptation to reconfigure the one

Comic by Chris Harden

in the show control system. But what works at home may not work in the park.

On one memorable installation I spent nearly an hour trying to debug a system that incorporated a PC, then discovered the maintenance crew had reconfigured it to install a copy of the game "Doom".

So PCs are only used for very specific applications in show control, where the system is largely insensitive to the hardware configuration of the PC itself.

Real Time Show Control

There are two fundamentally different types of show control systems: real time show control and scripted show control. The hardware used in these two types of systems is as different as the applications.

Real time show control systems are used for animatronic and other tightly synchronized shows. They are programmed by artists working at consoles covered with knobs. These "programmers" overdub the animation data in much the same way that a musician lays down tracks for a song. The animatronic figure's elbow is on one track, the shoulder on another, and so on.

Real time show control systems are almost always recorders. They capture the changing data from the knobs on the fly and then reproduce it over and over again until the artist is satisfied with result.

Real time show control systems aren't just for animatronic control. They are also used for many live stage shows, or anywhere that a sequence of events must happen rapidly. They make it easy to tweak the timing interactively.

Photo by Jim Tsorlinis

Scripted Show Control

Scripted show control systems are used when you need to control the show using... drumroll please... a script.

Remember our theater example? The preshow doors, curtains, house lights, and all the rest? That was a typical application for a scripted show controller.

Scripted show controllers tend to be black boxes that live in equipment racks for decades, silently repeating the same sequence of actions over and over. Most people wouldn't even recognize them as computers, although inside that's what they are.

Scripted show controllers are designed so that they can't easily be fiddled with. Once the timing of the show is worked out, the Art Director doesn't want some enterprising maintenance person changing it. There is typically no GUI (Graphical User Interface) on a scripted show controller, just a small status display.

Scripted show controllers are programmed using a spreadsheet that looks, quite literally, like a script, with show times in the first column and actions to the right. If you'd like to see what the front end for a scripted show controller looks like, you can download a free copy of WinScript from www.alcorn.com.

Alcorn McBride V16X Show Controller

Let's look at some of the other stuff we might encounter as we stumble around the construction site that is our themed attraction.

People Counters

Have you ever wondered why the theater hostess always says, "Please move all way to the end of the row, filling every available seat"? It's because she knows exactly how many seats there are in the theater, and she knows how many guests there are in the audience. And they're the same number.

Is the hostess really that good at counting? Or is there some trick involved here?

There is a trick. It's called a People Counter, and it's usually provided by the Show Control Engineer.

As you entered the preshow you passed through a narrow space where an infrared beam counted you. The results are displayed near her podium. As each person enters, the count goes up. The People Counter is pretty clever. It never counts handbags as an extra person, and it even gets most of the children, unless they're plastered to their mother's hip.

Perhaps while you waited for the show a few people decided they would rather catch an early lunch, or skip the show entirely and simply Exit Through Retail. As those people departed, the people counter subtracted them from the total guest count.

When the People Counter reaches the theater capacity, no more guests are allowed into the preshow area. This is accomplished by either a velvet rope or a full body block, depending upon how determined the guests are.

So when those theater entrance doors open, the hostess does indeed know that she has one—and only one—seat available for every guest.

Sometimes the People Counter is tightly integrated with the show system. In Epcot's American Adventure it can control the show cycle time, allowing more "fill/spill" time for heavily attended shows than nearly empty ones (as if there were such a thing!).

People Counter

Countdown Clocks

Along with the People Counter, nearly every theater show also has a Countdown Clock. A Countdown Clock is a sort of people pacifier. It lets guests know how long it will be until the next show begins, and gives them something to stare at blankly while they stand in the air-conditioned preshow, recovering from heatstroke.

Some Countdown Clocks can be set to different amounts of time, depending upon how often the show is running. Others always start at the same time, but guests are not allowed into the preshow until there is that amount of time left until the next main show.

The challenge for the Show Control Engineer comes when the Countdown Clock must be themed. In simple attractions the Countdown Clock may be nothing more than an LCD timer. But that doesn't fit in too well with old West theming. Or a medieval castle. So if the Art Director wants your countdown clock to be a grandfather clock—or an hourglass—get ready for a challenge. If he wants a sundial, it's time for a new Art Director.

Countdown Clock

Synchronization

One of the biggest responsibilities of the Show Control Engineer is synchronizing everything in the attraction. Audio must be synchronized to video. Animation must be synchronized to audio. Each scene on a ride may be synchronized to the vehicle motion. Special effects must synchronize with sound effects. And in some rides—ones conceived by particularly demonic Art Directors—music must run continuously throughout the entire ride, but loop through different lengths in different scenes.

The Show Control Engineer makes all of this work together. He or she typically accomplishes this with elaborate scripts running in a show controller that receives cues from ride vehicles and issues commands to all the other equipment.

The Show Control System is like a giant switchboard, and the Show Control Engineer is a frenzied operator, patching inputs to outputs in response to the demands of a hundred different clients.

Special Effects

The Special Effects Designer really keeps the Show Control Engineer on his or her toes. Special effects incorporate just about any technology you can imagine: mechanics, electronics, high voltage, pneumatics (air), hydraulics (nasty smelling, caustic oil), projection, lighting, and even cryogenics (cold stuff, like liquid nitrogen—don't touch!)

Speaking of cryogenics, no, Walt Disney isn't frozen inside of Sleeping Beauty's castle. There is a basketball hoop in the Matterhorn, though.

But I digress.

Many of these special effects require local control boxes. A control box is a sort of industrial strength local interface that allows maintenance to service the effect when the attraction isn't operating. For example, a liquid nitrogen effect may incorporate an insulated tank (like a giant thermos) that must be refilled every evening. This can be a complex process, requiring pipes to be purged of air and then pre-chilled so the liquid nitrogen won't evaporate. Safety mechanisms are needed to prevent the maintenance guy from accidentally turning himself into Mr. Freeze.

Such control boxes are usually designed by the Show Control Engineer. They provide local control and monitoring, and often include local safety interlocks. Control boxes usually have a "Hand/ Off/Auto" switch. When the switch is in the "Hand" (or "Manual") position, buttons on the box control the effect. When the switch is in the "Auto" position the main show controller can issue commands. And when the switch is in the "Off" position the effect is disabled.

Catastrophe Canyon Photo by Stain Marylight

Monitoring

When we talk about Show Control Engineering it's easy to forget that control is only half the job. In nearly every controlled system, monitoring is also needed. That's because the control signals that move stuff, turn on lights, play audio, or trigger effects all expect a response.

Any well-designed show control system makes sure this response occurs. If it doesn't, the show controller takes one or more actions.

It could:

- 🖫 Shut down the controlled equipment if there's any likelihood of damage
- 🖫 Notify Maintenance the equipment needs to be serviced
- 🖫 Activate alternate or backup equipment, if available

The better-designed systems respond quite intelligently to such problems as they arise.

In a well-designed theme park, show controllers in individual attractions collect this information and relay it to a central monitoring point such as Epcot Central, or Legoland's administrative offices.

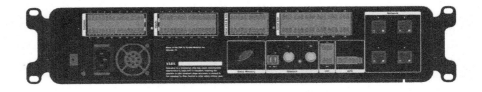

Alcorn McBride V16X Monitor I/O

Pressure Mats, Etc.

There's nothing that ruins a theme park engineer's day quite like squishing a guest. And it's amazing what lengths some guests will go to in order to get squished. Standing up on a roller coaster? Climbing out of a log boat while shooting through a flume? Crawling over the lap bar and jumping between vehicles? These don't sound like good ideas to me. But you'd be amazed how many guests will try them.

A lot of these unsafe activities fall in the domain of the Ride Control Engineer. (A few of them are governed only by God.) But the Show Control Engineer also tries to lengthen the lives of guests foolish enough to tangle with moving equipment.

If you climb out of a ride, chances are good that you'll step onto a pressure mat. The show control system will sense this and deactivate all action equipment in the area. In all likelihood the ride system will also shut down. And a lot of disappointed guests will stare angrily at you.

Pressure Pad

Break beams can also provide a measure of safety. If the guest climbs out of the queue area and into a space with moving set pieces, he's likely to break an infrared beam on the way. This will deactivate the set pieces and cause burly uniformed men to reunite him with his mother.

When the topography is too complicated for pressure mats, digital imaging techniques can be used. Video surveillance systems are now sophisticated enough that they can sense motion in portions of the picture where there should be none. If something moves where it shouldn't, the show control system shuts things down.

Things That Go Boom

The Show Control Engineer must sometimes control things that explode.

Explode intentionally, I mean.

What would a stunt show be without a few spectacular explosions? Most stunt shows involve pyrotechnics (fireworks), natural gas or propane, and big things that fall down. And, of course, it's not exciting unless there's an actor nearby.

Most stunt shows are only semi-automated. A Technical Director sits at the front of the audience, firing the effects as the stunt performers do the show. But Technical Directors aren't perfect. The stunt performers get very annoyed if the Technical Director sets them on fire. Stunt performers may be accustomed to having their eyebrows singed off, but they're rather fond of their major extremities.

So the Show Control Engineer comes to the Technical Director's rescue, and interposes a level of safety between the Technical Director's finger and that giant gas explosion.

For these applications the Show Control Engineer must design something more like a ride control system than a show control system. Fault tolerance, redundancy, and other ride control techniques are used, and the system must be subjected to a complete Failure Modes and Effects Analysis. These topics will be discussed soon.

Babelsberg StunT Show Photo by Dirk Brechmann

Design for Maintenance

If it ain't broke, don't fix it. But if it is broke, it better be fixable in one night.

Theme parks hate having inoperable attractions. So **Mean Time Between Failure (MTBF)** and **Mean Time To Repair (MTTR)** are important topics. The Show Control Engineer can help, by designing systems that are easy to diagnose and easy to repair. A well-designed show control system includes status displays that report problems, and special diagnostic modes that test the equipment, either automatically or manually.

As a Show Control Engineer, if you design a system like that you'll be the maintenance guys' best buddy. They might even take you out for a beer.

But if your system just sits there when it's broken, without giving a hint as to the cause... well do you see that bucket of tar and feathers over there...?

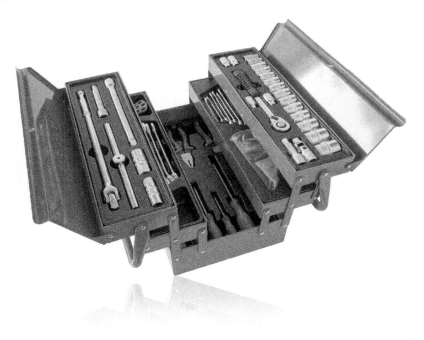

Photo By Jean Van Der Meulen

Ride Control

The easiest types of vehicles to control are ones that don't go very fast. Boat rides, for example.

Most boat rides are propelled simply by the motion of water around the Ride. Pumps aimed in the direction of travel maintain a constant flow, and the boats... well, they go with the flow.

With most boat rides it doesn't even matter if one boat bumps into another one, with one exception. It's very dangerous for boats to collide in the load and unload area. This can knock down people who are standing. To prevent this there is generally a "boat stop" just before the unload area.. It prevents boats from entering the unload area when there's already one there. Other boats back up around the Ride, and the boat stop lets one boat at time advance so the passengers can disembark.

Add a waterfall to your boat ride (the kind of waterfall the boats go down, not the kind that goes into the boats) and things start to get complicated. Now you need a boat stop at the top of the waterfall to make sure the boat can't rush down the waterfall and collide with a slow-moving boat at the bottom.

Add enough boat stops and enough monitoring and you've got what is known as a "block zone" control system. In a block zone control system the ride is divided into zones. These zones may be short or long. The smallest possible zone is at least as long as your vehicle—a boat in this case.

In a block zone system, a zone is considered occupied if there is a vehicle in it, and unoccupied when there's no vehicle. In order to guarantee that no vehicle can ever hit another, there must always be at least one unoccupied zone between any two occupied ones. A control mechanism—a boat stop, or simply turning off the propulsion system in a powered vehicle—stops any vehicle at the zone boundary until the zone ahead is unoccupied.

Block zone systems may be used in rides as simple as a boat ride, or as complex as a bobsled ride. (A bobsled ride is one like the Disneyland's Matterhorn or Space Mountain, with individual, high-speed vehicles moving independently around a track, all at the same time.)

While it might seem pretty simple to control a boat ride, it's often even simpler to control a ride—even a high-speed ride—that has only one vehicle. Surprisingly, roller coasters often fall into this category. On a roller coaster, only some rudimentary speed control and the ability to stop the roller coaster in the unload area are required. Beyond that, what goes up... must come down.

Photo By Mariya

Another easy type of ride to control is an Omnimover. These are rides like Disneyland's Haunted Mansion where all of the vehicles are attached to a single moving chain. Obviously vehicles cannot hit one another. In a way, there is only one vehicle; it's just very long.

Some rides are entirely controlled manually. A steam train is an example of this sort of ride. Usually attractions such as these don't even need Ride Control Engineers.

On the flip side, there are rides that go nowhere but require complex Ride Control Engineering. Simulators, for example, are extremely complex to control.

Most safety issues with simulators relate to the movement of guests—not the ride platforms. A sudden movement of the simulator during load or unload could injure guests. The ride system must hold the simulator in a quiescent state while guests are exiting or entering. This includes any elevators used to position ramps at the entrance and exit. Any movement of the simulator while the elevators are in place could cause serious damage.

During the ride, multiple sensors ensure each guest remains secured in his or her seat. Any "escape" will cause an Emergency Stop.

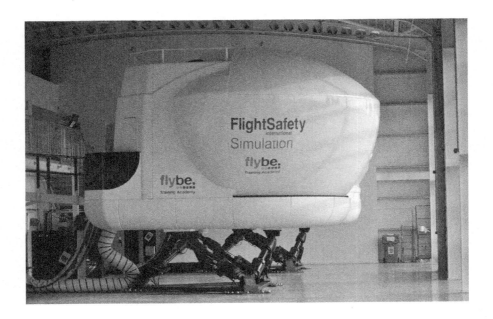

(One alarming story details the adventures of a guest who chose not to leave the exit ramp of a simulator before the exit doors closed. As the elevated ramp retracted he found himself in an uncomfortably small space through which the simulator violently swooped for five minutes as he cowered in terror. I'm not sure if this is an illustration that well-trained operations personnel are the ultimate safety system, or that there should have been pressure mats on the ramp. Or perhaps it's just evolution by natural selection.)

Ride Safety

If there's one single thing that distinguishes Ride Control Engineers from all of the other engineers who work on theme parks, it's safety. Ride Control Engineers live, breathe, and eat safety. They talk about safety with their friends when they go out to lunch. They talk about safety with their families on the weekend. They talk about safety when they're playing golf. Now that I think about it, Ride Control Engineers can be really tedious.

But if you meet a Ride Control Engineer who's not concerned about safety, run the other direction as fast as you can. Because when you get onto a theme park ride, you put yourself in the Ride Control Engineer's sweaty hands.

Murphy's law says, "Anything that can go wrong will go wrong."

In Ride Control Engineering, "Anything that can go wrong will go wrong if we let it. But we can't let it."

So the life of the Ride Control Engineer is one of constant vigilance—and fairly high stress.

There are many ways the Ride Control Engineer guarantees your safety. He has many tools at his disposal. This section presents just a few of the elements that Ride Control Engineers use to make your trip a safe one.

Computers vs. PLCs

Do you have a PC?

Of course you do.

Has it ever crashed?

If you said "no" with a straight face, you'd make a good politician.

So how would you like to be on a ride that depended upon your PC for its safety system?

Me neither.

That's why PCs don't make very good ride control computers. And yet you see them all the time, connected to simulators, monitoring safety systems, cycling vehicles. How is that possible?

We said that PCs made poor show control systems. But do you remember where PCs were useful in show control? One of the spots was for operator interface: status displays, statistics, cycle times, and so on.

PCs find their use in ride control systems in many of the same functions. They're good at displaying a lot of information in a small amount of space and allowing the user to tailor that information to their needs. But PCs are also used to do the actual control function of some rides.

Is this just foolhardy? Well, in some cases, yes. Simple attractions such as mall-based simulators are often controlled by a single PC. The reasoning is that if the PC fails the simulator will stop. And if the PC goes nuts, it can always be unplugged.

R-i-i-i-ght.

This sounds very logical at first, but it is actually a fairly flawed piece of thinking. I've seen simulator setups where a single failure of the PC output could cause the simulator to jump several feet. During the load or unload cycle this could be extremely hazardous.

A true Ride Control Engineer wouldn't do this, and you won't find such a system in a theme park.

Theme parks tend to use PLCs (Programmable Logic Controllers) for ride controllers. PLCs are industrial controllers designed for factory automation. They are comprised of extremely reliable hardware, but can be difficult to program, because they are primarily intended for simple automation tasks such as running a canning line. Nevertheless, they are the controller of choice for most attractions.

Allen Bradley PLC

Still, you say, I want to use a PC. Well, there is a way that PCs are used in ride control to directly control rides.

The incentive is simple. PCs are cheap and fast. They're easy to program. So they provide the most straightforward way to control complex new attractions.

Redundancy can make this safe.

Redundancy & Voting

Redundancy means more than one of something. If something has a one in a million chance of failing, it's regarded as fairly dangerous in theme parks. This may surprise you, but in a theme park it doesn't take long for something to happen one million times. Particularly if it's happening several times a minute.

If we now take two of those something's and require them both to agree for the action to take place, we'd like to believe that the chances of failure are now on the order of one million times one million. Unfortunately, that proves to be wishful thinking for most failure modes.

For example, if the failure is due to an error in our software, and the same software is loaded into both computers, then the chances of failure are still exactly the same as with one computer!

On the other hand, if the failure is a hardware failure (or a software failure caused by some hardware failure such as a memory fault) then using two computers does increase our reliability.

The simplest way to use redundant computers is to insist that the two computers agree at all times. If the computers ever disagree, then the system shuts down.

This technique works great for systems where shutting down is always safe. But it doesn't solve the problem in a system where shutting down the computers is, in itself, a dangerous thing to do. In systems like these we need to use yet another computer.

By using three computers to control a ride, we can create an extremely robust environment. For the ride to operate normally, all three computers must agree. If one of the computers disagrees, the two remaining computers are assumed to be correct. (The chances of the two computers agreeing and being wrong are about 1 in a gazillion unless the software is extremely poorly written.)

The advantage of using three computers is that when one disagrees, the other two can bring the ride to a safe shutdown while ignoring the odd man out. This means that trains can complete their ride cycles, and the guests be returned to the unload station. This is much safer than bringing the ride to an immediate stop and evacuating guests from it. (Ride spaces are not easy to walk through, and not all guests are physically able to navigate the evacuation routes, so this is a far safer solution.)

Triple redundant computer systems are only used in the most advanced rides, because they are expensive. But they're an ideal solution for dangerous rides when absolute safety is required.

Single and Multiple Point Failures

Ride Control Engineers spend more of their time analyzing the ride control system than they do designing it. The major thing they're looking for are failure points.

A failure point is anything in the system—hardware, software, mechanics—that if broken will cause an unsafe situation.

There are four general types of failures:

Single Point, Undetectable—This type of failure, when it occurs, is catastrophic. Imagine the brake pedal falling off of your car; You wouldn't notice until you tried to step on the brake, and then it would be too late.

Single Point, Detectable—Catastrophe can sometimes, but not always be averted in the case of this type of failure. Suppose there is a warning light that indicates the brake pedal has fallen off. It only helps you if you aren't in imminent need of the brakes.

Multiple Point, Undetectable—Catastrophe can sometimes, but not always be averted in the case of this type of failure. Suppose there are two brake pedals. If one falls off you might have time to use the other one. But what if the other one fell off undetected last month?

Multiple Point, Detectable—This is the only acceptable type of failure. Suppose there are two brake pedals and two warning lights. If one pedal falls off, you know that you should use the other one to stop the car at a service station and get it fixed.

It's easy to identify failure points in a system. It's much more difficult to correct them. Here's a simple example:

Most rides have brakes on the track, not on the vehicle. They operate against the vehicle, to slow it down or bring it to a stop. For example, suppose you have a gravity ride such as a bobsled ride. The vehicles are stopped in the station by an air brake. The brake is activated by an electronic output. If the brake is not applied, the vehicle can crash into the rear of the preceding vehicle, leading to an unsafe situation, injury, bad press, lawsuits, and general unpleasantness.

How can the brake fail?

It can fail mechanically: perhaps one-half of it falls off. It can fail pneumatically: the air hose falls off. It can fail electronically: the output fails to activate. It can fail in software: the computer fails to turn it on.

How can we make this system safe? Let's try this:

The maintenance staff inspects the brake every evening to make sure that it isn't damaged. They also inspect the hose to make sure it's securely attached and not cracked. Because they cannot inspect the electronic output we decide to add a second one. If either of the outputs comes on, the brake will be activated. Finally, we test our software under all known conditions to make sure it always activates the brake at the correct time.

Does that all sounds reasonable?

In fact, it's a disaster waiting to happen.

The fact that the brake is undamaged in the middle of the night says nothing about its condition by the next afternoon. The same applies to the air hose. And the fact that our software worked perfectly during test doesn't mean a thing. What if it's somehow altered, either intentionally or unintentionally?

Worst of all is the idea of adding a second electronic output in parallel with the first. This is what is known in the trade as a "multiple point undetectable failure". That's a big no-no.

Suppose one of the outputs fails. How would we know?

We wouldn't.

Which means we're now back in exactly the same situation we were when we had only a single output. In short, two outputs are no better than a single output from a safety standpoint. Unless the outputs are monitored.

If we monitor the outputs, things get much better. Now, when the first output fails, we immediately know that we're only one more failure away from an unsafe system. This allows us to unload the ride using our remaining output, and then repair the failed one before we load any new guests.

Similar monitoring could ensure that air pressure was always available at the pneumatic hose. But that doesn't solve our pneumatic problem. The pneumatic actuator itself may still fail to apply the brake. To improve the safety of pneumatic brakes, they require air pressure to RELEASE, not to BRAKE.

But this still doesn't fix our problem. Because the air valve may fail in the on position, causing the brake to be commanded to be released all the time.

In fact there is only one solution to this pneumatic problem and the associated mechanical problem of a damaged brake:

Take a good look at the track in the load area the next time you board a gravity ride. There are two separate brakes.

Cycle Testing

Why don't the two brakes have the same problem that the two electronic outputs had?

The answer is: cycle testing. Each time the brakes are applied, each is tested to make sure it mechanically closes on the vehicle. If it doesn't ,the failure is handled the same way the failure of the electronic output was handled. The ride is brought to a stop gracefully until repairs can be made.

(Notice that the entire foregoing discussion of safety relates to rides with multiple vehicles. This whole category of safety concerns disappears for rides with a single vehicle. Now you see why there are so many roller coasters with a single train.)

Photo By Luis Ramírez

Cycle testing finds application in other areas, too. It is the only technique that can validate a safety mechanism that is not normally used.

Suppose there is a track switch on our gravity ride. A track switch directs vehicles around the ride or into the maintenance area. The ride can only be operated normally if the track switch is in the position that directs vehicles around the ride. This is determined using a sensor. But what if this sensor is stuck on?

This sensor must be cycle tested. We must validate not only that it goes on when we command the track switch into the normal position, but also that it goes off when we command the track switch into the maintenance position. Seeing this transition guarantees that the sensor works, every time we move the track.

Surveillance

Break beams are another staple of Ride Control Engineering. They're particularly useful on boat rides for detecting the position of the boats. As with other sensors, they must be cycle tested. And on boat rides they must be located so that splashing water does not interfere with their functionality.

Brake beams are also useful for identifying situations that aren't supposed to happen: vehicles moving out of the ride envelope, for example. Unfortunately at this point it's somewhat like closing the barn doors after the horse has already escaped. Still, it's better for the system to know of the failure and shut down what it can, rather than continue to operate, oblivious to the problem.

Video surveillance is another barn door technique. Around most theme park rides are infrared cameras. (And yes, they can see you making out, even in the dark.) Somewhere a ride operator is watching several monitors, each usually with a four-way split screen display. These show him the activity of guests all around the ride. And while he may enjoy watching you make out, he definitely won't be amused by your standing up. Speakers throughout the ride allow him to issue warnings; if guests don't comply he can shut down the ride. It's a long walk to the exit amidst the glares of your fellow passengers.

Photo By Jürgen Jester

Pressure mats are reserved for the real idiots: the guy who manages to get completely out of the ride vehicle. Anyone who's ever seen the chains and gear and rotating tires that make theme park rides go has trouble imagining who would be stupid enough to do this. But people do. The pressure mats are a last-ditch attempt to shut down the ride before the guest pays a visit to that giant meat grinder in the sky.

Audio

Theme park audio equipment is very different from consumer or even commercial audio equipment. That's because it needs to run flawlessly and almost continuously for decades. There can be no rebooting of a frozen computer, no crashing of a hard disk.

On the other hand, the media almost never needs to be updated. As a result, theme park audio equipment generally isn't PC-based, and has no moving parts at all.

Alcorn McBride Digital Binloop

Audio Sources

There was a time when almost every audio source in a theme park used magnetic tape. Those days are gone. Now the audio sources have almost no moving parts. Sounds are stored in solid state memory and played back digitally.

While today MP3 is everywhere, it's not something you find in a theme park. And theme parks use far more audio tracks than you can imagine. A modern attraction can easily have more than 100 tracks of audio.

These tracks are typically stored as completely uncompressed audio such as WAV files, with sample sizes of 16 or even 24 bits, and playback rates as high at 192K samples per second.

If you multiply 24bits/sample times 192K samples/second, that's a lot of bits flying past! So Ethernet or similar cables are usually how these signals are routed, often all the way through to the final amplifier.

In addition to the complex audio requirements inside the attractions, there are also often hundreds of channels of BGM (background music) throughout a park.

And modern rides often have dozens or even hundreds of channels of audio on-board the vehicles, which often need to be mixed together in complex ways, to synchronize with off-board sound, and even adapt to each seat's individual speakers.

Alcorn McBride RidePlayer

Signal Processing

In a perfect world we wouldn't need to process signals. They'd emerge from the audio player exactly the way we needed them, be amplified and distributed to speakers. But in the real world things don't always sound the same in the theater or ride space as they did in the sound studio. Signal processing equipment allows the A/V Engineer to tweak things until they're just right.

Until recently nearly all signal processing consisted of rows of knobs or sliders that performed parametric or graphical equalization. Those are fancy sounding words for essentially a giant tone control.

The problem with knobs is that well-intentioned maintenance people can readjust the audio characteristics if they don't like the way it sounds. This tend to annoy A/V Engineers who've spent weeks adjusting things just the way they want them. So a form of equalizer was developed that had no knobs or sliders. It was programmed using a computer. Once the computer was removed voila! No fiddling allowed.

In the past few years there has been an almost wholesale move to digital signal processors (DSPs). DSPs are specialized computers that are very fast at mathematical operations. They treat all audio as a series of numbers, and can manipulate those numbers in whatever way they are programmed. This means that a single DSP is able to adjust the bass, treble and other frequency bands, and add delay, reverb, echo or even noise filtering.

This is extremely useful equipment to place in the audio distribution chain, because it allows nearly any contingency to be handled. Without DSPs, all sorts of other specialized equipment— such as switchers, delay generators, noise filters, or limiters—might be needed in order to get the show space to sound right. The DSP can play all these roles, without taking up extra rack space or adding cost.

Perhaps you've adjusted your home or car stereo for one song, and then found it didn't sound as good on another song. The same sort of thing can happen in a theme park. So digital signal processors are not only programmable, their programming can be changed "on the fly". This means that the show control system can send commands to the digital signal processor to adjust any characteristic of the sound at any time during the show.

Some of the signal processing requirements in a show are similar to those you might encounter at home. Others aren't. For example, your home probably doesn't have any rooms that are 200 feet long. (If it does, you might want to consider opening a bowling alley.) But 200 feet is a modest sized space in a theme park.

Why does the size of the space matter? Amazingly, it's because of the speed of sound.

In everyday life we tend to assume that things like light and sound travel from one spot to another instantaneously. This is actually a pretty good assumption for light. Unless you're trying to hold a conversation with astronauts on the moon, you're not aware of the delay that occurs over long distances for things—such as radio waves—that travel at the speed of light.

But sound is a lot slower than light. We use this fact all the time to measure how far away lightning is. We know that it takes about five seconds for sound to travel a mile, so by counting the time between seeing the lightning and hearing the thunder we can gauge the distance.

It turns out that even in a relatively small space—such as a 200 foot long theater—it takes a significant amount of time for sound to travel from the front to the rear. About one-fourth of a second, in fact. That doesn't sound like very much, but if you see someone's lips moving and then hear what they're saying, completely out of sync, it's quite annoying.

This effect can partially be eliminated by synchronizing the sound to the movie for a seat exactly in the center of the theater. That way in a 200 ft. theater no one hears the sound off by more than 1/8 of a second. But that still can be annoying. It becomes an even bigger problem in larger theaters.

In a theme park—or even your neighborhood Cineplex—this effect is compensated for by distributing the sound to different parts of the theater at different times. The sound is actually played back before the action occurs on the screen. Speakers aimed at people farthest away reproduce the sound first. The sound that goes to speakers near the front of the audience is delayed. The result is that everyone hears the soundtrack more or less in synchronization with the images.

Alcorn McBride A/V Binloop Rear Panel

Amplification

Once we have the sound modified just the way we want it, we need to make it a lot louder. The signals that roam around in the audio processing cabinet are tiny. It takes a lot more to drive the speakers in the theater.

This increase is accomplished exactly the same way as in your home or car: with amplifiers. But theme parks use a lot more amplifiers than you have in your home or car unless you drive one of those cars that pulls up next to me at the intersection throbbing so loud I have to readjust my rear view mirror.

It's not uncommon for a large theater in a theme park to require thousands of watts of amplification. (In this book I explain what's what, but not what's a watt.) Also, the amplifiers used in theme parks tend to be extremely high quality. After all, once we spend all that time and money getting the sound just the way we want it, we're not likely to scrimp on the power bill.

Crown High-Power Amplifier

Speakers

There are as many types of speakers as there are opinions about which speakers sound best. And that's a bunch. There are woofers, tweeters, electrostatics, horns, drivers, Venturi ports, bass reflex systems... you get the idea. Whatever I say about speakers will be obsolete in two minutes and disagreed with by half of the A/V Engineers on any given attraction anyway. Suffice it to say the ultimate goal of the attraction's speakers is to deliver undistorted sound to every guest.

There are two basic techniques for distributing sound to speakers. You're familiar with one from your home stereo. Using this technique the amplifiers expect to see a particular load on each audio channel. This load is measured in resistance, and expressed in Ohms. Your home speakers are probably eight Ohms, and your home amplifier is designed to drive either 4 or 8 Ohms on each of its outputs.

THX

Klipsch KPT-684 Dual 18" Subwoofer

Most theme park speakers work this way as well. But there's another type of distribution used in theme parks and other public places, including your neighborhood shopping mall or department store. It's called 70-volt audio distribution. Those flat, round speakers you see mounted to ceiling tiles usually use this scheme.

70-volt audio distribution requires a transformer on the amplifier output, and another one at each speaker. You might wonder why anyone would want to go to the trouble of using a system that requires so many extra parts. The reason is that using a 70-volt distribution system we can hook many speakers to a single amplifier output, and with very cheap wire. And the contractor can wire the speakers in the ceiling willy-nilly. And with some contractors, willy-nilly is as good as it gets.

While 70-volt speaker systems are perfectly adequate for most background music applications, they don't provide sufficient volume or audio quality for theaters.

PA

PA stands for Public Address, another of the A/V Engineer's responsibilities. Most themed attractions have PA stations throughout the ride. These can page into individual areas or throughout the entire attraction. The system is generally integrated with a park-wide PA system as well. These days the PA system is usually controlled by the DSP, or Digital Signal Processor. It sounds simple at first, but the system must be able to interpret any one of a hundred buttons and route the live audio from that mike to the preassigned areas associated with that button while allowing prerecorded audio to continue playing in other areas. Think of those annoying "Now boarding rows 25 and higher only" pages at Gate 73 and you get the idea.

In some themed attractions the A/V Engineer also installs intercom systems at the Operator Control Consoles so that the operators can make dates with each other between shows.

Alcorn McBride Paging Station

Video

Historically, video in the United States has operated at 60Hz, while much of the rest of the world is at 50Hz.

Most theme park video is high-definition (HD) or better. There are many different resolutions of high-definition video, but the most popular home standard produces a picture 1920 pixels wide by 1080 lines high.

Unfortunately, the high-definition standards were developed for broadcast use, where frequency spectrum costs money. So the standards committee elected to severely compress the video signal in order to save bandwidth. The result is a picture that is not as sharp as it could be. Theme parks installing high-definition video players usually select equipment that operates at many times the broadcast frequency. They also use resolutions far beyond the standards. Nearly all new theme parks are equipped with digital video players that use either memory or hard disks for media storage.

HDMI is a digital distribution format popular with consumers. Because of cable length limitations it isn't always a good choice in large installations, but it's cheap.

SDI is one of the best (and most expensive) video distribution formats available.

There are also many new video distribution formats for ultra high-definition formats including DisplayPort and video over various forms of Cat-5 cable.

Uncompressed Video Formats

With the advent of 4K consumer products, ultra high-definition video is becoming more affordable, although even 4K pictures contain only about 2,000 lines of resolution, so in theme parks multiple images are often seamed together.

Most high-definition video standards use compression, so the video is encoded after it is shot, before it can be streamed or distributed in other formats.

But that's not the case in theme parks, where the quality of giant video projections must be far better than you'd encounter at home or even in a digital cinema. So theme parks strive for uncompressed video, which offers a perfect picture with no compression artifacts. The challenge is that it requires enormous amounts of storage and extremely high transmission rates.

Most theme parks now use memory-based video playback devices such as the Alcorn McBride Digital Binloop line. And since the video is uncompressed, as many images may be perfectly seamed together as needed, making an 8K or even 16K output no problem. This allows large theaters such as Soarin' or Flight of Passage to display ultra realistic images.

These devices have no moving parts. The Mean Time Between Failure (MTBF) of this new equipment is measured not in days or months, but in decades.

Alcorn McBride A/V Binloop Uncompressed

Video Display

A well-designed video processing cabinet in the equipment room will provide a local monitor, so that the output of the video players can be checked without walking into the show space.

Out in the guest area the spaces are large, so any monitors need to be, too. The big screen monitor that looks so huge in your living room seems a lot smaller in a preshow area. Theme parks need large displays, and lots of them, often connected side by side seamlessly.

Video projection is also popular, and the latest ultra high resolution laser projectors can create very large images at very high resolutions. It is also possible to almost seamless combine these projections at the images using a bit of feather and overlap. That's how a show such as The Simpsons a Universal Orlando can fill such a large dome with a very sharp image.

Hanging the Rear Projection Screen
in American Adventure

Lighting

We all know a little bit about lighting, because we've all changed a light bulb at one time or another. Or stuck our finger in a light socket. So we can imagine some of the challenges that Lighting Designers face: wattage requirements, heat dissipation, electrical code requirements. These details must be worked out years before the attraction opens, but so far it seems pretty familiar. But there are differences.

In your house a switch controls most of your lighting. We flip a switch and electricity is connected to the light bulb. It comes on at full brightness. End of story.

Very little of the lighting in a theme park attraction is controlled that way—often only the fluorescent worklights, which are only used when the park is closed.

In your dining room perhaps you have a dimmer switch that lets you adjust the brightness of the chandelier over your dining table—

for those romantic dinners for two. Or so people can't see the crayon marks on the wall.

The lighting in our themed attraction is controlled using a computerized version of that dimmer switch.

The lighting controller is often a lighting board, although this is an expensive choice. Lighting designers like to use a lighting board during test and adjust, because it allows them to tweak each fixture to the perfect brightness. So the lighting board often becomes a permanent part of the installation. But a better way to design the lighting control system is to transfer the settings from the lighting board into a dedicated piece of equipment designed for permanent installations. This prevents the settings from being fiddled with later.

The connection between the lighting controller and the dimmer cabinets is usually a signal called DMX (DMX stand for Digital Multiplex, which means a bunch of digital values interleaved).

Photo By Klaus P. Rausch

There are a few types of lighting that are difficult or impossible to dim. Florescent lighting is notoriously problematical. Although not much conventional florescent lighting is used in theme parks, a specialized form of it called ultraviolet or "black light" is. "Black lights" make florescent paint glow spectacularly in the dark. There are few themed attractions that don't make some use of this technique. Most "black lights" can't be connected to dimmer cabinets, and must be turned on in the morning, and off in the evening. This is still done by computers, through a computer-controlled switch.

Perhaps you've seen moving spotlights at a rock concert. These fixtures are complex to control, usually requiring multiple channels of DMX. Different channels control the brightness, the color, and the motion of the fixture.

Special Effects

Fog is a staple of the Special Effects Designer. It swirls about mysteriously, masking the sins of the Set Designer. It also creates a fluid surface for the Lighting Designer to play with. And it looks cool.

Theme park fog comes in a couple of varieties. The most popular is atomized water droplets. In other words, real fog. It's created by releasing water onto rapidly vibrating metal plates, which break it into tiny droplets that hang in the air just like the real thing. The advantage of this water-based fog is it's clean and harmless. The disadvantage is it makes your attraction mildew.

Another disadvantage of water-based fog is that it takes a while for it to build up. It's not suitable for attractions were a sudden cloud is needed in a short burst.

For these applications liquid nitrogen (LN2) is used. When released into the air, LN2 is so cold it instantly vaporizes, creating a dense white cloud of condensation that quickly dissipates. This sudden burst of fog is useful to mask the secret workings of attractions. For example, at Universal Studios a blast of LN2 disguised the ascent of the Back To The Future vehicle into the domed theater.

Air is seventy percent nitrogen, so liquid nitrogen—as long as it doesn't get on you—seems fairly harmless. And it is. But a problem occurs when a lot of LN2 is released in a confined space. The nitrogen can easily displace the air, and before you know it you're breathing an atmosphere containing no oxygen. Thud.

There's nothing like an unconscious guest to give an attraction a bad reputation. So attractions that use a lot of liquid nitrogen incorporate nitrogen sensors, to make sure the air remains... well, air-like.

Fog Machine

Another popular tool of the Special Effects Designer is projection. That's an extremely useful way to expand the rear wall of the attraction into the infinite. Spacious skies, amber fields of waving grain, purple mountains caressed by majestic clouds, you get the idea. When projected onto a painted scrim from the rear they're also used for smaller effects, such as burning torches or shimmering seas.

The Special Effects Designer's bag of tricks is huge, so I'll just mention one more favorite: smell cannons. These are devices that release a small amount of scent, usually triggered by the passing of a ride vehicle. Popular scents include hot rocks, roses and stinkbugs.

Well, that's about it for the Electronic Engineering and related roles in theme park design. If one of these captured your fancy then you might want to consider coming not just an Imagineer, but that rarest of all beings, The Engineer Imagineer!

Epilogue

Mickey's Ten Commandments

So why do theme parks still fascinate us, in this world where even the neighborhood mall is themed?

It's because of the care with which theme parks are designed, the attention to detail, and the impassioned commitment of theme park designers to transport their guests into an alternate reality.

Every theme park designer should know what's been done in the past. That's why former Walt Disney Imagineering President Marty Sklar developed his ten guidelines for theme park design. They're called Mickey's Ten Commandments.

Here they are, in slightly paraphrased form:

- Know your audience. Don't bore people, talk down to them or lose them by assuming they know what you know.

- Wear your guest's shoes. Insist designers, staff, and your board members experience your facility as visitors as often as possible.

- Organize the flow of people and ideas. Use good story telling techniques, tell good stories not lectures, lay out your exhibit with a clear logic.

- Create a weenie. Lead visitors from one area to another by creating visual magnets and giving visitors rewards for making the journey.

- Communicate with visual literacy. Make good use of all the non-verbal ways of communication: color, shape, form and texture.

- Avoid overload. Resist the temptation to tell too much, to have too many objects. Don't force people to swallow more than they can digest. Try to stimulate and provide guidance to those who want more.

- Tell one story at a time. If you have a lot of information, divide it into distinct, logical, organized stories. People can absorb and retain information more clearly if the path to the next concept is clear and logical.

- Avoid contradiction. Clear institutional identity helps give you the competitive edge. The public needs to know who you are and what differentiates you from other institutions they may have seen.

- For every ounce of treatment provide a ton of fun. How do you woo people from all other temptations? Give people plenty of opportunity to enjoy themselves by emphasizing ways that let people participate in the experience and by making your environment rich and appealing to all the senses.

- Keep it up. Never underestimate the importance of cleanliness and routine maintenance. People expect to get a good show every time.

I hope you enjoyed this book. We've covered a lot. I've tried to give you a sample of every delicacy from the buffet of theme park design, but I'm sure we've overlooked many tasty treats.

If your career plans carry you into this busy and exciting field, drop me a line someday. In this business everyone has a story to tell. I'd love to hear yours.

Perhaps we may even run into each other behind the scenes...

...most likely at 2 o'clock in the morning on opening day.

Now please gather your experiences, passions and inspirations and step into the vehicle. Your ride is just beginning.

About the Author

Steve Alcorn is an entrepreneur, engineer, inventor, author and teacher best known for his involvement in the theme park industry. In 1982 he joined Walt Disney Imagineering (then known as WED Enterprises) as a consultant, where he worked on the electronic systems for Epcot Center. During his time with Imagineering he designed show control systems for The American Adventure, wrote the operating system used in the park-wide monitoring system, and became Imagineering's first Systems Engineer.

In 1986 he founded Alcorn McBride Inc. The company's show control, audio, video and lighting equipment is used in most major theme park attractions around the world.

Mr. Alcorn is the author of many novels and non-fiction books, including How to Fix Your Novel, and Building a Better Mouse: The Story of the Electronic Imagineers who Designed Epcot. His books are available through amazon.com. You can find out more about all of his books at themeperks.com

During the past few decades, Mr. Alcorn has also helped more than 50,000 aspiring authors structure their novels through his online classes, offered at over 1500 universities and colleges worldwide. Find out more at writingacademy.com.

If you enjoyed this book, you might also enjoy the class he teaches in Theme Park Engineering at imagineeringclass.com. It lets you design your own attraction and receive feedback from Mr. Alcorn.

Steve Alcorn